W9-AZG-611

THE GENTLE JUNGLE

THE GENTLE JUNGLE

Toni Ringo Helfer

Brigham Young University Press

Photo on cover by Arthur Wien of Wild Art.

Library of Congress cataloging in Publication Data

Helfer, Toni Ringo, 1940–
The gentle jungle.

SUMMARY: Describes the history and techniques of wild animal affection training as perfected by Ralph Helfer.
1. Animals, Training of. [1. Animals–Training] I. Title.
GV1829.H4 636.08'8 80-10275
ISBN 0-8425-1790-1

Brigham Young University Press, Provo, Utah 84602

Printed in the United States of America
4/80

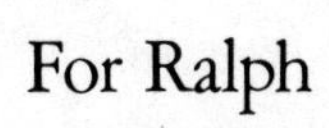
For Ralph

Among the noblest of the land,
Though he may count himself the least,
That man I honor and revere
Who without favor, without fear,
In the great city dares to stand
The friend of every friendless beast.

–Henry Wadsworth Longfellow

Contents

Foreword

I have known Toni and Ralph Helfer for years and through them have experienced some of the most touching, exciting, and the funniest episodes in my life–everything from bewildered lions turning to me for protection from dancing, singing schoolchildren to the sad death of a 22-foot pregnant python in Nanyuki, Kenya. If you read this book, you will understand some of these experiences. These two people have accomplished more with affection training of wild animals than anyone I know.

Years after our initial meeting in Africa, I visited Ralph and Toni at Africa USA in Saugus, California, and there I was even more amazed at the rapport the two of them had with their wildlife. Toni is an absolutely fearless human being who is totally loved by every animal she has ever touched. She is undaunted by any endeavor or undertaking. A remarkable painter, naturalist, zoologist, conservationist, and now author, Toni Helfer has the courage and the curiosity we all should have. For the sake of the world I wish Ralph and Toni a long, rich, and productive life.

William Holden

Introduction: Two Redheaded Men

Years after the events I have described in this book, while I was attending a conservation convention in Texas, I took the afternoon off and went to the nearby zoo. Zoos have never been among my favorite haunts, but I believe for this particular time in history they are necessary. I have only wished that with their allotted funds, each would specialize according to location, making breeding areas as close to an animal's natural environ ment as is humanly possible, spending their money on quality housing for a few, eliminating cages, and establishing colonies of well-displayed, free-roaming clans (to the extent they may be free).

Each zoo would initiate a special study of perhaps ten varieties of species and end this two-of-everything nonsense. In this way the animals would adapt to specific, well-planned surroundings that elaborated on areas, one would hope, not designed for a gaping public but for the comfort and reproduction of that particular breed. Adaptation would center around that animal's habitat, whether it be trees or burroughs, caves or rock quarries, savannahs or forests. One would have to visit zoological gardens throughout the United States to see a multitude of species. Perhaps the zoo, if conceived in this fashion, would become the highlight of every city. Such a visual composition would complement its inhabitants and would probably increase tourism to some extent.

On this particular summer Sunday as I disturbed the confining silence and walked from cage to cage, feeling the value of life and experiencing the freedom denied those within these restricted bounds, I came before a cold, dirty, grey enclosure, barred in front by heavy, black iron poles, insuring those behind against ever breaking free, creating forever a maximum security. I sat upon a gaily painted bench, made comfortable to please man. And here I stayed, nauseated in a way that an empathetic animal can be, in pain for living things.

Hurting, in deep emotional disorder, I weighed the suffering of this innocent world. Children collected to make mock of the dignified orangutans in the cell I was facing. They threw peanuts and popcorn at them and contorted their faces and stuck out their tongues and thumped their chests, strutting back and forth as though they were superior little men and women before inferior beings. When they had made their sport and had their humiliating fun, a teacher herded them to the cage beyond, and so the naughty game carried on, in contempt of the great value and excellence of the lives within. The instructor occasionally joined in the amusement and encouraged the insensitive imitation. I thought their education sadly lacking as their ignorance paraded by.

The orangs pressed their faces tightly to the prison bars and extended their arms in a friendly way as though inviting the children to stay, perhaps hoping to play. And as I observed that those two "old men of the woods" did not understand their condemned condition, I began to cry. Now they looked at me, and one stuck out his tongue in hopes of lightening my load; the other gave me the "raspberry," and I stood and drew near to the railing, filled with appreciation that they so kindly had changed the subject.

I began now to laugh and say, "You silly old things," and from my throat emitted squealing sounds as I spoke in their native tongue. And they began to respond. Their house became a lively, social room of uncontrolled excitement. Then I really looked at them. The one large ape climbed the poles before me and held out both arms to me, and as though he knew me, patted his head, then covered his eyes. My heart began to pound, and I said, "Genghis, where's your brain?" And he tapped his temple with his finger. And so that I would not mistake his identity, I asked, "Give me a smile." Both red-headed men showed their broadest grins. I yelled, "Dear God, has it come to this?" and over the rail I leaped (knowing the risk) to embrace my long lost friends, Hannibal and Genghis Khan. Muscular arms wrapped around me, arms I had known many years ago. We hugged and touched and spoke of things long past, of joys in a world that had been better then.

I have never known such agony as when I met again those two crazy men, whom I had known as boys. I had taught them when they were students of mine, but I had never told them of zoos and isolation and boredom in a place where no one would recognize them. When they had been taken from us at the dissolution of a business partnership, I had no idea what would happen to them. They had waited so long for a caress and a kind word. I held their hands and questioned their separation from me, then questioned *Sick Society* as well as myself. And not until a horrified keeper came and parted our company did I stop.

As they led me away as though I were a criminal, the orangs' screams filled the space, and I broke that day, my heart as well as my mind. Try though I did, offering even my soul to get them back, the authorities dis-compassionately preferred to let them be. The city and inhumane laws prevented man from befriending animals such as these, and Hannibal and Genghis Khan may still be waiting for my return. It was not a question of money (for I offered far above the usual price) but one of inhumanity. And I will never again hold my Hannibal or Genghis Khan, nor they me.

Experiences such as this one—and others of a deeply personal nature—became a collective force compelling me to write this book. My story is about animals responding to the love of humans and about the humans who were profoundly affected by that response.

1
In the Spring of the Bears

Stone walls do not a prison make
Nor iron bars a cage.

–Richard Lovelace

1

After I married Ralph, I rode King many times into the hills above our ranch and sat astride him through long, satisfied musings, enjoying the bustling activity taking place below. My life in the veiled morning sun seemed near perfection. I was in love with absolutely everything and felt infinitely privileged to be born to such a time and to belong in such a place. My days, though laden with growing responsibility, were free from the suffocating thoughts the rest of the world seemed to be dying from. The atmosphere of my life was characterized by a special kind of placidity, overflowing with happiness and real worth. I'm not one of those people who has known all along what she would do with her life. I have enjoyed accomplishing many things, but I did not come into this world with singular, channeled thoughts as Ralph had. Through me swept enormous pride in my husband, for he had known from the very first that his life would be among the animals.

The irony is that such tranquility had come to me out of utter chaos.

Everything had started peacefully enough. It was July 3, 1961. My sister Stevie, our father Scotty, and I were beginning our vacation in esthetic Sequoia National Park. Every summer as far back as I can remember we had camped at Atwell Mill. When Stevie and I were very young we'd mark the days off with self-inflicted excitement, building toward a crescendo of vocal explosion on the day we left home and began to climb the treacherous hairpin turns just outside Three Rivers.

The mountains were alive with self-consuming and self-reviving vegetation–lush shaggy grass, glistening delicate ferns, brilliant wildflowers drenched in sunlight. The earth beneath the thick carpet of pine needles sparkled with copper iridescence; the sky shone indigo blue, and the air was breathless. Faint whisperings sailed through this consecrated place, to infinity and beyond.

Amid this splendor I held my sister's head while she threw up by the side of the road. I felt myself turning a shade of pea green. Nothing made me quite so sick as watching someone else retch. What a way to begin a holiday.

I dabbed my hankie in the cold water trough and wiped her face, trying not to look down.

"You know, for someone who always gets carsick I don't know why you sat in the back seat."

"Uuuughhhh," she moaned as I led her back to the car and deposited her in the front seat.

"How are you feeling, honey?" Scotty asked.

"Real crappy," she groaned.

"Stevie! Do you have to use that word? It's so vulgar." I winced.

"It's a gift," said she.

"Well, it seems to me it's a family curse! With all your brains you could surely think of another way in which to express yourself," I criticized.

"Girls, will you please stop fighting. We're on holiday, remember?"

"This is true. I'll forgive her, but only because she's sick."

"Forgive me? You'll forgive me? How generous of you."

"Girls, girls!"

The sun was just sinking as we turned into the old Atwell campground. A thin rim of deep crimson clung to the edge of elevated silhouettes, then slowly drowned in earth.

We came to a stop at the far end of the dirt road, just on a meadow's edge.

While we unloaded our gear and began to set up camp, a troup of boy scouts, three sites down, lit a blazing fire and began singing.

"Oh, my darling Clementine" rang out through the hollow. Stevie began to sing along.

I interrupted. "You know your voice gives a new meaning to the word *music.* You're off key as usual. Tune in next time."

"You're just jealous," she snapped back.

"I thought you were sick."

"I am, and just to prove how sick I really am, I'm going to consent to listen to one of your warped ghost stories, breaking forever the dull monotony of my short but well-lived life. You may begin."

"How thoughtful."

Here among the Sequoia pines, I began a ghostly tale similar to the ones I used to tell Stevie when she was younger, paralyzing her with delicious fear.

"Ahhh, how splendid! The mountain hours are far enough into the night and the mood deep enough that a macabre tale, one suggesting ghastly terror, would seem to have found its time," I said with sinister appeal, in an endeavor to begin scaring the wits out of my demented sister. I eased into a gruesome story, shaping it with dramatics to prolong the agony, and I unexpectedly began to scare myself. Scotty, our father, having long since left to recapture his youth with the scouts, was barely within earshot. The night suddenly began to close in on us. My voice began to crack, when, from behind the boulder at the back of our camp, came an eerie rustling of brush and the soft crunch of snapping twigs. Neither Stevie nor I could speak. We were encased by the dark, and we felt sheer fright.

"What's that?" I quavered finally.

"What's what?" she answered.

"That," I said.

"Ohhh. . . . *that.* Ahh, why don't you go see, Toni?"

"Me! What a bizarre idea. Why don't *you* go see, oh Fearless One?"

"Shhh. . . . Whatever it is, it's breathing," she whispered. "Shhhh. . . . listen."

"Oh, it's probably just a squirrel, for heaven's sake. We're so silly."

"Here, take the flashlight and go have a look." She gave me a shove in the direction of the sound. Her eyes were about to explode from her face.

"Not on your life, Chicken Little. If I go, you go with me." I grabbed her sweaty hand as she said, "Permit me to sacrifice myself."

Much to my regret, I was first. I shone the light with a shaking hand about the black woods. Not a thing did I see until I came to a protruding ridge in the rock and peered around at a great, fuzzy bear peering back at me! Did we run, did we flee, did we fly! Three more startled beings I have rarely seen dispersing in three opposite directions. With a bark of amazement the bear disappeared into the dark. Stevie went straight up the boulder. To this day neither of us can figure out how she did it. I was in the car and had the door locked before I realized my sister was still outside. I drove to where she sat, and in one of those continuous uncontrolled movements, she came sliding down, colliding with the door on the passenger side. With a frantic gesture she tried to open it, converting energy into rage as she pulled and tugged. I leisurely bent over and lifted the lock. She fell rigidly in and was immobile.

When we woke from an uncomfortable sleep in the safe car, it was to the smell of toast and trout cooking over an open fire. We descended upon the morning, stiffened into a 45-degree angle. The veiled sun limbered our bones, and a wash in ice-cold river water nearly broke them. We swooped down upon the meal like locusts devouring a crop. The fish had been donated by the nearby scouts, to whom we were very grateful.

It was a glorious day, and our interlude with the bear was now a laughable thing. Scotty was off chopping wood and giving birth to pathetic bird calls. Stevie and I cleaned up camp with less than enthusiastic zeal.

"A woman's work is never done," she cried, dusting the bark on a nearby tree.

My sister, four years younger than I, was a high-spirited, wild sixteen-year-old with a marvelous quick wit and a spontaneous sense of humor. She had been born with academic wisdom, but being brighter than one's instructors led to sheer boredom and a need for diversion. My mother often had well-founded anxiety when the school nurse called inquiring as to Stevie's health, when in fact she was often off seeking refuge with others who had an undisciplined rage to live.

Suddenly Stevie came running for me, grabbing my arm and knocking me about.

Keeping my cool, as always—having to preserve my big-sister status—I announced with some detachment, "Something seems to be wrong here."

"Look!" she yelled, pointing across to the empty campsite beyond. "A bear!"

Angry flies and bees scattered, as one by one the bear pulled the garbage cans over and ravaged their contents, collecting her morning meal.

"I bet she's the same one we saw last night. She doesn't look quite so fierce, nor so large now does she, Toni?"

"Nope."

It was a lovely picturesque moment, one that seemed to typify the spirit of the gentle mountains.

Our tranquil wilderness scene was interrupted as the distant dust of an oncoming ranger's truck hazed the view.

Stevie nudged me. "Look–Ranger," she said primitively.

"Aha, you do have extraordinary eyesight," I answered perkily.

"Maybe he has a friend." Her eyes reflected a romantic fantasy. "This could lead to the big one, Kid," she said, elbowing me in the side.

"It's past your bedtime, Dear," I said.

"What are you talking about? It's nine a.m.!"

"I know, I know, but take a look in the mirror. That fright last night must have had a damaging effect on your whole system. You look awful."

"I do?"

"Yes, dreadful. You wouldn't want anyone to see you like that, now would you? After all, we owe it to our public to keep the Ringo Sisters' image as high as possible."

"We do?"

The truck came to a startling stop directly in front of us. My pulse quickened with mounting inquisitive tension as, without haste, the door to the truck was opened and out jumped a uniformed fiftyish man (it could happen to anyone). Stevie whispered, "You want him? Okay, he's yours."

He had a rifle in his hand, and he dashed around to the front of the truck, dropped to his knee, took a quick aim, and–before we could register what was about to happen–he shot the bear!

"Oh, dear God, no!"

She flailed about in agony for a tormented moment, then fell on her side and lay still. The forest ranger stood, walked over to the animal, pulled a knife from its leather sheath, lifted the bear's great head, bent low, and slit her throat.

Horrified, Stevie and I clung to one another, staring at the gruesome sight, not believing what we had seen.

Recovering from shock, we began to shout at him.

"This is a national park! A wildlife sanctuary!"

"No hunting! No shooting!"

"You killed her!"

But our pain-filled outburst fell on deaf ears as the man began to drag the slaughtered beast by her hind legs, leaving behind a trail of innocent blood. The bees and the flies began to collect again, but this time it was not around the garbage cans. Calm and unemotional, the man heaved and tugged until he had loaded the corpse into the back of his truck. Then he turned and shouted, "Lady, this is a public campground. We have a right to shoot these beasts when they make a nuisance of themselves and endanger the lives of vacationers." Covered in red death, he drove off and left us bewildered, enveloped in a cloud of dust. So much for the voice in the wilderness. Jet streams of gleaming light filtered through the dark green roof of stately trees above our heads, but nothing could recapture our unrealistic youth of fifteen minutes past.

"Let's get out of here. This place is sick!" I said. Scotty, who had returned when he heard the shot, tried to calm our wild imagination. But we had never seen through hate before. (Years later I was told by a well-known conservationist that each spring before the tourist season starts, a pit is dug and packed with tempting food to lure unsuspecting bears, who, when they begin to fill their empty stomachs after a long winter's rest, are promptly slaughtered.)

As we threw the last sleeping bag into the car and looked about for anything we may have left behind other than idealistic adolescence, an unfamiliar yowl caught our attention. The sound came from the vicinity of the murder, prompting in us a mild noxious sensation as we headed for the dismal spot. The woebegone wails were now above our heads. I looked straight up into the seductivity of a Bierstadt painting, where, perched high on a flimsy limb, shook tiny furballs–bear cubs–isolated and hanging on for dear life. So on top of everything else the forest ranger had killed Mama bear! Well, he wasn't going to get her babies, too–not on your life! We climbed that tree, put those cubs into our car, and shooting stars sailed across a patriotic sky as we drove down from Hell's Mountain on that Fourth of July.

2

I was thankful for one thing on our return home: the bears slept like cherubs all the way down the mountain.

At this point in my life I was attending UCLA Extension and working as a junior executive at the May Company. I had a two-bedroom apartment on Rossmore Street. It was the only building on the street with a canopy out front, and I thought that leaned toward class distinction. It was a nice, warm building filled with the exuberance of aspiring actors.

Having driven me to my place, Scotty now left the problem of the two cubs with me. Picking them up by the back of the neck, he carried them up the stairs to my room, and, as I opened the door with my key, dropped them inside, and murmured these never-to-be-forgotten words, "Good luck." And he and Stevie left me in the presence of ignorance. "Thanks a lot, Dad."

It was dark, and the light switch was above the sofa across the room. There wasn't a sound. I tiptoed, groping my way in the lamp's direction, when one of the cubs let out a blood-curdling scream and ascended my stunned body the way a telephone repairman wearing cleats climbs a wooden pole. He pulled himself up with what felt like grappling hooks, puncturing me every bit of the way. To my shock he reached my startled face and, looking directly into my popping eyes, he screamed again. Ah, ah, my heart! Having completely intimidated me, he slid down my frame, shredding me painfully, stopping only at my boots, where he bit them and shook his fierce head as though he would tear me limb from limb. I dragged him with my foot to the lamp and switched it on.

Then came the sound of breaking glass. I stumbled into my bedroom and fell in front of my dresser as two hundred dollars worth of perfume fell before me. The second cub dived from the dresser, landing directly on my displaced back, then bounced with athletic zeal out the door and toward the kitchen. I dragged my battered self up with the help of collapsing drapes, pulling the cub on my leg with me, debating whether to save myself or the apartment first. I tried one last time to loosen the thing from its death grip. Looking fiercer than ever, it yelled up at me, then uttered filthy little sounds.

"Okay, okay. Relax, it's yours." And my added weight and I, walking a great deal like Quasimodo, the hunchback of Notre Dame, abandoned the bedroom and slammed the door behind us.

Seeing how much fun the other cub was having destroying my kitchen, he let loose of my numb limb and joined in. Things began to fly everywhere; they had raked their way to the counter top and were hastily devouring wood, wall paper, food, drink–absolutely everything they came in contact with. They began to ice skate on the tile, grabbing, as they slid by, anything they could demolish. The pile now on the floor was growing rapidly. I couldn't understand it. Disney films were never like this.

I am proud to say that in the face of disaster I'm all knowing. I gathered the last of my energy, glared with increasing disgust at the cubs, took a deep breath, turned off the lights, and escaped downstairs to my girlfriend's apartment, where I spent the remainder of a sleepless night.

Early the next morning I raced upstairs, heart pounding. All was quiet within. Slowly I opened the door, but I was not prepared for such a sight. If I were to describe the end of the world, its devastation would pale in comparison. Nothing had remained untouched. The stuffing had been torn out of the sofa and pictures pulled from the walls. White-flour paw prints were everywhere. Little was left of the kitchen. It was the most incredible mess I have ever seen. And sound asleep in a corner, snuggled angelically together, lay the demons of the night, exhausted from their exploits in captivity.

After a brief appraisal of the damage, I weighed my alternatives, considering changing my name and leaving the country, or going underground. What would my landlord do? Put me in jail? Sue me? I shut the door—with me on the outside—and stood, catatonic. An actor who lived down the hall came by. He had never so much as looked at me before.

"What's wrong, Toni?" (I didn't know he knew my name.) I couldn't speak, not one intelligible word. So I opened the door to the catastrophe. He looked in and to my surprise was not driven back by shock. Instead, he acted with instinctive perception. When you've lived in Hollywood for a while, I suppose you get used to anything.

"Toni, you are not to worry about a thing." He gave me a brotherly pat on the head. Astonished by his confidence, I came back to consciousness.

"I'm afraid I don't understand, Sal."

"Well, it just so happens," he confided, "I'm making a film with the greatest animal man of all time, and would you believe I have his phone number in my wallet?" He reached for his billfold and pulled forth a card. "Here!" and he handed me the crumpled business card, beaming. Come on, let's go call him."

We tiptoed through the wreckage to the only intact piece of furniture. Although off the hook, the phone still worked. One cub raised a weary head, gave a half bark, and tossed us an ingratiating smile before falling back into blissful sleep.

"How do you pronounce this man's last name?" I asked.

"H-e-l-f-e-r. Ralph *Helfer,*" Sal emphasized.

"Wait a minute," I said before dialing. "Why am I calling him?"

"Because unless you *want* to spend more of these wonderful moments (and he made a great sweeping motion with his arm about the room), you will give those little monsters to Ralph before things get any worse!"

"How right you are," I agreed and dialed the number. "Where is this place?" I whispered as the phone rang.

"Saugus." It sounded like a social disease, not the name of a town. (Ah, yes, the creeping, crawling Saugus has got her, poor thing.)

"Ralph supplies the film industry with exotic animals," Sal explained. "He's the creator of 'Affection Training.' There, Nature's Haven." He pointed to the name on the card. "That's the name of his ranch." (Nature's haven? Affection training? It had to be a nudist colony!)

3

Observing a sunrise on a crisp California morning from the slopes of Yosemite gives rise to a high spiritual element, but that anyone would live in the bottom of a canyon in Saugus was more than I could understand.

The ranch was somewhat like an oasis in the dry, brittle desert and that was an exhilaration; but around it not a tree was to be found. One rode in the naked air and swam in waves of heat and everything was bare and stripped and scorched. There was no fantasy here, except, I knew, sometimes in winter when occasional snowflakes fell. I am a mountain person, and the land without trees hurts my eyes.

Nature's Haven was wedged in a canyon on the Agua Dulce Mountains. On a cool summer day it was 115° in the shade, and the nights were freezing. It was 1,500 acres of incredible Vasquez Rock formation. Great, flat, sharp, upthrusting rocks that had been flung from the heavens in disc fashion had landed with their filed edges pointing toward the nemesis of sky. Everything was arranged with colorless taste. The famous bandit Vasquez was said to have buried his wealth of treasure somewhere in this natural citadel.

I drove down the old Borax road and turned onto a dirt drive where the sign read *Nature's Haven.* My borrowed car coughed from its consumption of dust.

It was beginning to grow dark, but a rosy glow colored the bland topography, temporarily jewelling the mystic shapes and crowning them with a flush of eloquent light. I had carried with me enough food for an army and sent a steady stream of goodies shooting to the back seat to occupy the cubs while I drove.

As I pulled up to the trailer I'd been instructed to stop in front of, I honked several times. When the porch light went on, I abandoned the car to the playfulness reigning within.

As the screen door opened, it had the familiar squeak of that of a mountain store. Only the sound of the little bell was missing.

It was then that an extraordinary, ageless man emerged and came down those steps toward me. An explosion of personality filled the air, and a rush of adrenalin swept through my body.

He walked toward me like an Indian, toes first. His body was so tight I almost expected his muscles to burst through. And he was radiating pure animal magnetism. I tried with great difficulty to compose myself as he came near.

"Hello; welcome to Nature's Haven. I'm Ralph Helfer."

Ralph and leopard–pure animal magnetism.

"Hi there, I'm Toni Ringo."

"Well, well, what have we here?" He looked into the car window and chuckled in a contagious way.

"How on earth did you drive all that way with that avalanche falling around you?"

"I just kept feeding them." (Boy, does that sound dumb now.) He smiled and shook his head with good nature, and we opened the door (more intent on one another than on the car and its contents). The bears ejected themselves in a cycle of energetic, swift renewal.

The enormous German shepherd at Ralph's side went dashing after the hairy streaks, and only seconds went by before he returned with one of them dangling between his teeth. Ralph took hold of the bear by the back of the neck and deposited him in a nearby cage. When the dog returned carrying the other cub in the same fashion, I was awestruck. "Why, he's absolutely amazing!"

"Yes," he answered, putting the second escapee away. "Prince is the residing genius here, aren't you, Boy?" and he gave the dog a fond pat, to which the Shepherd responded with regal adoration. "Nature's Haven belongs to Prince; we're his guests, sharing equally in the things he likes best."

"I'm afraid I don't understand."

"No, of course, you don't. I'm sorry. Can I offer you some refreshment?" he suggested, changing the subject.

"No thank you; I really should be getting home."

"Wouldn't you like to see the animals?" he asked with a note of injured dignity.

"I thought you'd never ask."

We walked and talked that evening as though we'd known one another all our lives, and as he gave me a formal introduction to things I had never anticipated meeting I had the distinct feeling he was showing me his soul.

Mumbles and grumbles and groans of exaltation filled my ears as he stepped from one cage to another and spoke with intense affection and scratched each joyful inhabitant through the chain-link fence.

Huge tigers and bears reacted like kittens to his touch, and wolves and coyotes wagged from head to toe, anticipating their turn. These, to me, were new vibrations, new music, new sounds—a new harmony to which this man responded with the same tone—the very essence of his character.

We came to a stop before a long enclosed run and Ralph called out, "Hey you lazy moose, we've got company. Wake up."

One big amber eye opened as the animal's hind quarters slowly rose, long front legs stretching far out in front of him and huge paws kneading the ground beneath them. A yawn that would have put Jaws to shame exposed a mouth filled with sparkling dagger fangs. The lion now stood to his full 529-pound size. Then, with a sudden magical transformation from his primitive character, he began to act like a puppy, bounding with leaps and twists through the brisk evening air. Ralph opened the cage door and was greeted by a mighty forward thrust as Zamba sprang up with a muscular effort and placed both front feet gently upon the man's shoulders.

Impressed? Oh, no, I was not too impressed—only rendered speechless.

The lion now flopped down as though all four legs had been jerked from under him. He rolled around on his back like a playful kitten. To my amazement, Ralph dived on him and the two wrestled like deranged brothers. I found this display a traumatizing experience. I do not look at all attractive with my mouth hanging open, my eyes bulging, and my neck extended.

Ralph now stood and the lion stood with him, rubbing against his leg like the family dog. There was a tom-tom sound as Ralph slapped the big cat's sides. He threw his arms around the mammoth's black-maned neck and squeezed him in a big hug.

Wrestling like deranged brothers–Zamba and Ralph.

"Do you like lions?" he asked.

"Adore them," I answered in fear of my life.

"Would you like to come in and meet Zamba?"

"No!" I bleated.

"Well, maybe next time," he said as he stepped out and locked the door behind him. This was without a doubt the most incredible person I had ever come across in fiction or real life.

"I hope you won't get the wrong impression of me," he pleaded with a shy grin. "But you see, Zamba is my best friend; I could never pass him by."

"No, of course you couldn't," I said for lack of anything else to say.

We wound our way down neatly inlaid rock steps and entered an old but lived-in barn. It was here that I first met the fabled circus elephant known as Modoc. Her great crippled trunk draped around me, and I was overwhelmed, spellbound, enchanted—not to mention already hopelessly in love with the Animal Man. Among the ways to tame a wild bird of prey is not allowing him to sleep. So I spent the next few hours question ing Ralph. I was intrigued by what motivated this fascinating, mysterious person.

"Where are you from?"

"Chicago."

He sat down on one of Modoc's hay bales, leaned back against the wooden wall, and placed a piece of straw between his teeth. Prince lay at his feet. I knew he must be plagued constantly with questions, but I couldn't stop. Without hesitation or misgiving I continued my investigation.

"Have you always had this affinity with animals?"

"Always."

This was not going to be easy. Prince looked up at me with a kind of esoteric smile—far more human than dog—and with a tinge of delight, as though telling himself funny stories, he gave a half wag of his tail.

"What are you laughing at?" I was actually questioning a dog!

"You see, Prince, we're beginning to have an interesting effect on her."

Ralph watched me closely as I sat "lotus" next to Prince. I felt like a disciple about to subscribe to the teachings of a master. I still have that first-etched impression of Ralph's thick, tawny hair, worn in a crew cut, of skin richly bronzed by the desert sun. His nose was aquiline, and those blazing sapphire eyes were deep and penetrating. He had the qualities of a hawk. His was the face of a kind man, a good man, who had felt great love and joy and pain for all living things. In this high energy field I felt my temperature rise.

"Ahh,—ahh," I stuttered, trying to catch my breath, "won't you tell me what you meant when you said this was Prince's home and you were a guest here?"

"Yes. Yes, I would like you to know him as I have come to. He's to be greatly admired."

Bingo.

"Would you mind if I read you a little history I have written about

Prince?" he asked with a covering modesty. (I should have known a poet lived here as well.) I nodded with mounting eagerness.

"I'll be right back. Prince, you keep a protective eye on her." And with that he was gone.

He hadn't been gone two seconds when the elephant, swaying not ten feet away from where I sat in a dreamy, distant state, let out a bellow that not only deafened me, it splattered me against the wall, jerking me instantly out of my fantasyland.

I was on the verge of becoming hysterical as I snapped back to paranoid reality. I mean, what did I know about this guy? He could have made a habit of luring unsuspecting girls like me out to this remote isolated place, only to fatten us up and feed us to the lions. And his dog was probably in it with him. Dear God, they'll never find a trace of me. Even my bones will be gone.

The door swung open and Ralph appeared, papers in hand.

"Is anything wrong?" he asked.

"Oh no," I answered with a note of surprise.

"Then why are you pointing that pitchfork at me?"

"Oh, how silly of me. I was just tidying up around this hay pile."

"You're not afraid of Old Modoc are you?"

"Well, now that you mention it, I've never been alone with an elephant before. It's a little unnerving, isn't it?"

He smiled with kind eyes and looked fondly at the old elephant who, with her chin tucked, looked feebly back at him and rumbled. Theirs was a speechless communication. Modoc fanned her frayed ears and began to rock from side to side, swaying in an ancestral pace, like a glorious ancient ship riding on high seas, shadowed partially by the night. I was standing in the wings as Ralph went over and gave her enormous trunk a big hug. Then, patting her bulging sides, he said to me, "Come here." I felt a great compassion as I approached the massive grey body.

"Toni, Modoc is to be respected but never feared, and to be her friend you only have to say her name."

"Hello, Mo," I said, feeling a bit ridiculous at first.

"Modoc is the great old majesty of the animal stage, one of the proud legends of circus history," he announced as though considering it an honor to know her. Turning to me, he held out a hand and coaxed me into petting her. "One day," he promised warmly, "I'll teach you how to handle her."

The thought filled me with unexpected pleasure.

I followed dreamily as we went back to the hay bale. When Ralph sat, I crouched near his feet, and the dog sat at attention as if he were about to indulge himself in his own past. It seemed to me we were a perfect audience as Ralph began Prince's story:

With cool, limpid nights came the wail of the coyote and a sneak of marauding raiders, but none was so bold as the wild dog from those hills. Often, from a distance, I would watch as the large German Shepherd hunted small game to provide his day's meal. The dog was shy, as are all wild animals intent on survival. I could not get near him. His instinct told him man was a dangerous beast.

He appeared to have been a mountaineer in those canyons for some time. And at dusk when the coyotes began to yelp, I heard in the chorus a call like that of a wolf–a deep, low baying, sad in that it was alone.

I admired him with an inquisitive interest. What was this splendid canine doing here? Yet I respected his removed existence and never made my way toward him. Once or twice when I climbed high into the slanted Vasquez Rocks and looked about my canyon, my eyes met the fiery gaze of the dog posed in dignity across the valley on an opposite perilous wall. I realized that I was the intruder here. I waved and wished him well and hoped for the day when he would regard me as a friend.

One evening I was startled by a quiet shadow advancing toward the carnivores' houses. As I flashed a light to the path beyond, it fell upon the wild dog, who ripped and tore at a piece of meat protruding slightly from the jaguar's cage. Snarling, the dog kept the gentle jaguar cowering away from his supper. Such courage and endurance was worth recognition. As he pulled loose the meat, I called out "Bravo!" and he and his free meal turned with a start. The uninvited guest faded into the night while I replaced the bewildered animal's dinner.

Having bested the creature whose unfamiliar voice so disturbed him, the dog found renewed pleasure in raiding my cages and so abandoned the hunt. During the day, with a curiousity-filled observation, he lay beneath the desert silence, blending with the landscape, at the property's edge. He never barked, never wagged his tail, but watched with patient demeanor, unconcerned by my activity. I began to throw him bits of food, a bone, some meat.

> Weeks later, cautiously, the dog began to follow after me—growling, deep-throated, faithful to his stipulation, and independent. Mud-caked and weather-worn, he came with little consideration into my life of gentle wild animals. Now he lay his defiant head upon his enormous paws and watched with a sense of calm resolve, yet inner misery. I placed a bowl of water and a plate of food next to the army blanket where he occasionally slept with one eye open.
>
> How many evenings my horse Son and I rode in those canyons. Blue blood ran through Son's veins; he was a noble breed, proud and strong. Eager and adventuresome we'd go, into the wild dark smudges of night. And the dog ran ahead to lead the way. No longer could I call him wild dog. I knew him too well now. So I named him *Prince.* Prince loved the horse, for what is anything when you are alone? People die of such things, you know, and in his newfound happiness I felt my own.
>
> On weekends I often rode early in the day. I spied many deer, and the dog at my side would break and run to play; no calling on my part would stop him. I'd watch the frisky chase for miles, until they were long out of sight. He might be gone for days, a week, a night. Then back the vagabond would come, to settle in the pasture by the horse he'd come to know.
>
> Prince was now guarded by only thin fears. None of us had abetted his distrust of man. Slowly his set of wild values crumbled away. He began to accept the world with a new understanding, fighting his old ghosts, beast to beast. When love finally overcame him, the savageness vanished forever, and he became more than I had expected. He was the boyhood friend I had never had.

At the finish of the tale he lifted his eyes from the page to look at me.

I was wiping my nose on my sleeve and dabbing that same garment to my eyes. I swell when I cry, and I turn red. What could I say in such an awkward moment? "Do you happen to have a Kleenex?" I quavered.

He reached into his hip pocket and took out a handkerchief (I didn't think anyone carried those things anymore). He smiled and with a calloused hand tilted my chin toward him and wiped my seeping eyes. Then he *kissed* me on the forehead.

"Well, if I was a frog before, I'm a princess now," I managed to croak,

Prince and friend.

and the two of us began to laugh. (Talk about show business instinct.)

I drove back to L.A. a different person, fully aware I would never be the same again. But the weekend passed without a word from him. I survived those few days only by keeping that night alive in my thoughts. Then on Monday after school the phone rang.

"Could we have dinner together on Wednesday?" Ralph asked. "Yes, I think we can manage that," I gulped.

Away from his surroundings Ralph had a social shortcoming (one I would have to work on immediately)–a lack of clothes sense. It was a startling fact that he dressed anything but conventionally. That first night when he came to pick me up for dinner at Kelbo's, he was wearing a jacket with no shirt under it, shoes with no socks. From his neck hung an ivory bone so large that an elephant should have been attached to it. I don't know how he carried his head straight.

Being my mother's daughter, I, of course, had impeccable taste and was dressed elegantly in a black suit, white gloves, and pearls. We looked rather unusual together. In his hand was a copy of *The Prophet.* (Oh no–not that book again!)

As I glanced at our reflection in a window, my impression that we were a strange couple was intensified. But when he spoke, I saw again the beautiful soul, and I simply resolved that I would have to perform some delicate alterations to make the frame fit my picture. How could I know he was thinking the same of me?

We made our way out the front door, down the stairs, through the lobby, onto the street, and into the parked station wagon with the lion in the back seat......LION IN THE BACK SEAT!!!

Oh dear, I had so many little things to overcome.

5

In the weeks to follow I was hopelessly caught up in the spell of my newly discovered utopia. Although I wasn't at the Agua Dulce ranch more than a dozen times, my fondest memories are of those first afternoons when I walked with Ralph, Zamba, and Prince in the narrow canyon behind the compound. The lion took great delight in exploring the rocky ridges on his own, and although he enjoyed those times of running free, he preferred above all to walk by Ralph's side. Sometimes Ralph would climb on Zamba's back and ride ahead of Prince and me, and it seemed they were as happy in one another's company as two animals could be.

Zamba was without question an extension of Ralph, his alter ego. Ralph was Zamba's first person and Zamba was Ralph's first lion. Through that lion and their extraordinary friendship Ralph was allowed to experience life from the animal's sphere.

Zam, nature's abattoir, was Ralph's mentor in a Golden Mane.

"Toni, there's something I have to discuss with you."

His voice sounded serious. I waited. I was eager but doing well at not showing it.

"Zamba and I are leaving for Kenya, East Africa, in two weeks. We're going to work on a film based on Joseph Kessel's book, *The Lion.*

"You are?" I said with a wounded heart, trying not to show my disappointment. "I suppose it must be a high point in your career—I mean, taking lions to Africa and all. That's rather like taking coals to Newcastle, isn't it?"

He nodded his head.

"How long will you be gone?"

"Six months."

"Six months! Half a year? Why, that's a lifetime." I do not deny that with every breath he took I waited for him to include me, but he didn't.

"I'm very happy for you. I hope with all my heart that the film goes well. Just imagine, Africa. That's not a place anyone really goes to; it's a place we all dream of." I looked down in a last desperate effort not to cry.

"Toni, Toni, Toni."

I began nervously picking burrs out of Zamba's mane.

"Look at me, Toni."

"I can't."

"Oh yes you can." He grabbed me by the arm, pulled me against him, and said, "I'm in love with you."

The overscore of *Wuthering Heights* filled my romantic ears, along with Zamba's roars as he called out from the mountain's edge, across the moors, to the valley beyond.

Ralph doubling for Massai warrior in The Lion.

That last evening before Ralph left, we walked the desert path worn by the endless march of Nature and Ralph's perfect parade. Those canyon strolls, where eyes watched from above and all around, were now meaningful to me, and I absorbed this world in spiritual oneness, abolishing all the ugliness of the sphere on the other side in the world of man as he is today–transformed by society–out of touch with life. I wanted to keep this perfect place close to me–to protect and cherish it as one does all loved things.

It was our last stroll for six long months.

The Lion was to establish Ralph as one of the great pioneers in captive animal behaviorism. That film was the first time a full-grown African lion had ever worked with a small child. The beautiful triangle love story between a lion, a Massai warrior, and an eight-year-old child was not only controversial, but, until Ralph and Zamba walked into Sam Engel's office at Twentieth Century Fox, it had been considered impossible to shoot.

I did not visit Nature's Haven while Ralph was gone. Instead I buried myself in work and school. My youthful romantic heartache needed diversion.

Ralph's letters to me have become very important, for they convey more of this man and his sensitivity than I could ever communicate.

11/15/61

My Dearest Toni,

Time and time again, until the day when I shall leave no more, I will come home to Africa.

When Zamba and I alighted at Nairobi Airport, a transformation occurred in our character–so great that one might suspect we had survived passage through time. We arrived like a luminous speck of dust from high in the air and settled with brilliant incandescence–a metamorphosis.

As if this transformation had been arranged especially for me, I descended upon the quixotic town of Nairobi–a chivalric hero I, idealized by the adventure and extraordinary mystery of things to come. A lion not unlike Rosinante walked by my side. The Black Sea parted as Zam and I roamed the quaint foreign streets.

Our reverification brought about an air of profound respect when we entered as kings the swinging doors of the New Stanley Hotel. The waiting crowd divided into branchlike parts as we wandered the corridors of African heritage. We found

Zamba and Ralph roaming the streets of Nairobi.

many of the rooms in various stages of colonial decay.

After the press conference we were described as avant-garde, and the critics regarded us as leading in our field. The next week our photos were in every periodical and newspaper, and the country of Kenya was struck with curiosity as we became the object of much inquiry.

We found ourselves traveling a bumpy dirt road, occasionally blooming in white morning glories, winding its way past hillside Kikyu huts to Nanyuki's corrugated tin shops and the Brown invasion of Indian intrusion. The horror of Mau Mau's bloody wave still fills the distilled air, and the king's African rifles parade in stately columns with preoccupied countenance, guarding with whole regiments the camp that lies at the entrance to the Mount Kenya Safari Club.

Since I was a small boy, I had wanted to work with animals. Since then, too, I had wanted to go to Africa. Now here I am at 7,000 feet, in the romantic setting of Mt. Kenya, 12,000 miles from everything I had known, passing from an artificial deception to nature, obvious and unashamed.

I have climbed a mountain with luxurious bamboo forests. I have fished from her serpentine radiant streams, walked the sylvan glades past wooded gorges that give passage to a downpour of vertical water. In the early beam of morning I have seen the Bongo, the mountain's living gold. The Alpine moorlands I have crossed past giant senecios, lobelea, gladioli, orchids, and pericum alive with yellow blossoms and brilliant sunbirds–far more glorious than any dream. A hundred kinds of birds have I seen: turaco, francolin, thrush, hawk, eagle, owl, weaver–an intoxicating number of birds–and mammals large and small. I have looked about this altar crowned by twin peaks–Batian (17,058 ft.) and Nelion (17,022 ft.)–a mantle of white shadows sleeping. From the cold air of heaven they are covered with fine snow. I stand light in the atmosphere. The friendship of my mountain is sacred. I have seen God here.

12/1/61

Dearest Toni,

The silver sky is filled with an explosion of dark puffy clouds, which occasionally overflow into a damp vapor of weightless mist that gently sprays the endless miles of long, waving grass. Fields of air and dew hang between the earth and the sky, and only my feet are on the ground.

It was on this celestial day as I walked through the green life of Africa that I turned Zamba loose, on his home soil. The filtered sun set the world aglow, and silhouettes of elephants with gleaming ivory tusks walked the horizon in a straight line.

Zamba was loose–free to roam and run upon the plains where he was born and from where he had been kidnapped–a land of vast, inspired creation, untouched and becoming from the beginning of time.

After he had sniffed and explored and done his silly bouncing dance of joy, Zamba marked his territory, then lay a short distance from where I sat beneath an acacia tree. He lay in the fashion of a beautiful Kuhnert painting, atop a hyena burrough facing the softly blowing east wind, his regal mane flowing from his golden face.

I couldn't help thinking of the wrongs that had been inflicted upon the people and animals of Africa. Both had been stolen and sold into slavery, deprived of their heritage and stripped of tradition. (Where would Zamba be if he weren't with me?)

I could never bring all of Africa's children back here; it would be neither possible nor fair, for they would not survive. They have evolved beyond this time. They have been forced against their will to live within the progression of a white, uncivilized culture.

I can think of nothing more gratifying than to stay here forever with Zamba, to enter his world with him, but it would come to no good eventually. So we shall be satisfied for now to enjoy each moment of recovered paradise and contemplate a world that should have been.

Yet it would be my fondest dream to buy a farm on the slope of Mt. Kenya, and the torn roots would mend, and we'd walk in the blaze of the African sun and set everything right and atone for such inhumanity.

12/15/61

Dearest Toni,

How great the pleasure of wandering. I revive from a long sleep, and the evening's recital gives place to the dawning day's exploration. The air is delicious, exhilarating, inspired by the cold. I came upon my first elephant in the wild in the hours just before dawn in the Aberdare Mountains. Here in the luxuriant foliage of human depopulation, in a small clearing amidst imposing masses of metallic green, high up in the clouds he stood, and so large did he appear that I was dwarfed.

As far as he was concerned, my presence was no more or less than any exile who, on such an occasion, was made to feel insignificant by the appearance of the obvious ruler of the wilderness. I suffered from never having seen him before. And he, as do all fine spectacles, did his best to enhance that brief interval by wallowing in a thick bog of rich, soft mud to protect himself from the insect bites and the sun's blistering beam.

It was easy to imagine ten thousand years of elephants having done the same, as he dug his wealth of ivory into the brine, scooping and hollowing out a pool. Raising his trunk, he sniffed the east wind, and standing with shifting footfalls, he spread his ears fanwise and with an oscillating pace crushed the foliage beneath the weight of his great feet. Then, with an explosive sound, emitted as a result of something unknown to me, he spun and traversed the distance between our boundaries.

I, having nowhere to go, climbed the nearest tree with expert agility, only to find myself perched just below an olive baboon, who leaned down to gaze with disbelief and dismay at the strange, colorless, displaced body that clung so vigorously to his tree. Embowered in shade and rendered utterly helpless, I watched as the monarch fled into the green darkness below. On a purely creative impulse I stayed attached to my perch, thinking all the while what a fine tale the gaping figure above would have to tell.

From the corner of my eye I saw a soft movement of elephant grass tinged with a yellow rim. Slowly the ripple of grass edged toward the clearing, where a thick crop of buffalo began to disassemble, calling in discord as they went, a combination of tones suggesting unrelieved tension.

As the tall, jointed stems parted, a lustrous head of great beauty appeared, crouching low to the ground, ears pinned back, eyes blazing, occasionally twitching with anticipation, but waiting–patiently waiting. The buffs stamped their feet defiantly, looking about wildly, sniffed where not even a whisper of wind was near, and emitted guttural, threatening sounds. I was dripping wet, caught between alarm and expectation, transfixed by the drama.

Instinctively I twitched as the lioness catapulted with rigid muscles, and with increased speed focused on the kill. She was powerful and grand in her violence. From out of the bush came three others like her; and they worked as highly skilled sheep dogs do, singling out the prey, isolating him from the others, taunting him, as tawny fury slashed and tore and ripped and clung and bit. Weighted down by feline pounds, the brave warrior was pulled to his knees. And before sweet death came, they had torn open his flank. Though he fought courageously, their strategy won. Had his kind joined forces, he would not have been surrounded by the cats. But they had stood quietly, showing no affection, gazing on the flagrant injustice, gazing on, gazing on, even as I.

With weariness he fell and so died, and his friends retired into the bush whence they had come. The feast then began.

Vultures now sat, like me, in a nearby tree, and a black-backed jackal stood close by. A stream of fresh blood flowed into a crack of hard-baked mud, and a foul stench smothered the air as flies covered the carcass, and the lions, submerged in entrails, occasionally revived and loudly accosted the cat nearest them.

Not until two hours later did the lions drag their bellies away from the carcass beneath the canopy of vultures. They pointed their feet toward God and rolled from side to side in discomfort from having stuffed themselves. The gluttons moaned and groaned and cleaned their faces and fell fast asleep as vultures and jackals descended upon the leftovers.

I cautiously climbed down, and the sky began to rain as I made my way through the dense hedge of bamboo, down a trail of running water, through slush and clinging mud to the comfort and safety of my Landrover with Zamba's face painted on her doors. If Zamba had witnessed Africa, savage as I had seen her, would he have joined it? Somehow I doubted it. I would have had to teach him how to kill as I have taught him how to live.

Back at the Safari Club I bathed in Africa's red water, then dried by the fire Njoka had built for me. I set my mud-caked boots outside my door to be cleaned. The music from my tape recorder changed my mood to thoughts of you, and I was suddenly alone and wrote you of my day.

This is a place so perfect, so absolute, so flawless, where time chose its most beautiful moment to stand still—a wonderful lost land of refined souls, exquisite spirits, perfumed air—another world, finished as God would have it, imperfect as man would see it.

12/20/61
Dearest Toni,

As a long-distance Christmas present I'm sending you a page from my diary:

July 6, 1961

How penetrating these hours filled with happy satisfaction. The sun is down, this day has nearly passed. There is a deep happiness in me tonight. I signed the contract to film *The*

Lion. Zamba and I leave in early October for Africa. I have had a brief remission of melancholy, as the only lack in my life now is someone to share my adventures with.

At 8:00 p.m. a young girl came into my life. She climbed out of a very active car and stepped into the light of my front porch. Brushing back her long, honey-colored hair, straightening her blouse and tucking it into her shorts, she extended a gracious hand and introduced herself. "Hello, I'm Toni Ringo; I'm sorry I'm such a mess, but as you can see, I had no cage to put them in, and I've been fighting for the wheel the entire trip." She was the maximum in health and radiance, tall, tan, willowy, and ethereal. She didn't wear shoes, and when she smiled as she extended a friendly hand I thought, "Uh oh, here comes trouble."

"Mr. Helfer? You are Mr. Helfer, aren't you?"

My shaking limbs and I approached her car with tough bravado.

"Well, let's see what we have here?" I opened the door, and two brown bears jumped out. (That was a good way to impress her! Good grief, Helfer, she just told you they were loose.) Thank heaven Prince was with me. He grabbed the second cub on his way out by the back of the neck. I took hold as Prince chased the other bear and returned with it dangling from his jaws.

"Why, that dog's amazing," she said as I put the bears in a nearby cage.

"He certainly is," I replied, counting my blessings for having him there and saving me from disgrace.

After I put the cubs away I had to think of something to keep her for awhile. I offered her refreshments, but to my surprise she said she had to leave. I thought I was losing my touch. Then the animals announced themselves, allowing for the order of things.

"Wouldn't you like to take a walk around the compound and see the animals before you go?"

"I would love to," she replied.

I did my best to impress her. Whether I did or not remains to be seen. When she drove away I had a strong sense of loss. I didn't want her to go. I stared after her for a long, thoughtful while. This was indeed a special day.

Toni, these months in Africa have affected me so profoundly that I feel transformed. When I return home, it will be with a mind that sees all things differently, with a faith that has always centered on nature, and with a belief that I will pursue with even more devotion. I see you now as belonging only to me, and having become surprisingly territorial and primitively possessive, I seem to have lost all sense of never having been in love.

Will you marry me?

Forever and ever,
Ralph

2
The Helfer Chronicles

"Little children never give
pain to things that feel and live."

–Anonymous

1

To yesterday I gave myself, and I soared to the very gates of heaven, which opened just a crack. Therein I felt the sympathy of universal understanding–a quality of great kindness within the warmth of all good things. And when I call to mind my life among the animals, I shall be happy with such a memory. Now, from the course of many years, to my last faithful friend I bequeath these remembrances, so others may know that once there was a man of uncommon perception.

He is by all outward appearances a perfectly ordinary man; but my man sees the world through the amber eyes of a lion, and although his thoughts are with peace, his heart roars with a song of the wild, of golden manes blowing on moving air. My man sees the wind, and he speaks with the animals, and he is alarming in his implications. And when he walks, it is with the love of all things. So deep are his considerations that complete harmony distinguishes him.

In the following capsuled biography I hope to give you more than just a bit of history of someone who is still very much alive. I have tried to record as accurately as possible feelings as well as actual events that happened in Ralph's early years so that perhaps you, to a small extent, may come to appreciate him as I have.

But before we start this journey, I propose a toast: Here's to all children born of certain vision, to their strong desires for high achievement, to the quest and the fulfillment of their dreams, to those who keep up the fight and never tire. Here's to the child, to the boy, to the man of extraordinary insight. Here's to you, Ralph.

Ralph David Helfer was born April 9, 1931 in Chicago, Illinois. At the age of four he spoke of his love for animals in a rare and rooted way, illuminated with the bright intensity of a child's fantasy. Those youthful adventuresome daydreams were the foundation of a future reality.

Uncle Chan was head of the Helfer household. It was his large apartment that he, Ralph, his sister Sally, and their mother and father shared.

Uncle Chan was subject to unwholesome outbursts, but because they all lived together, a good deal of the man's character was overlooked. One might compare Uncle Chan to an austere ballet instructor, who, keeping within a narrow and specified discipline, makes his pupil perform his exercise under devout and imposing guidelines, forcing upon the pupil everything he had ever learned and harping on the way in which these things were done, demanding perfection and all the while, inside, believing it was for his pupil's good, not his own.

Chan received a kind of sadistic pleasure from shooting the Chicago sewer rats. Sometimes he acted like a psychopathic scientist from an experimental research institute. And he made his sensitive young nephew assist, forcing Ralph into setting up his grisly source of enjoyment.

Ralph told me the following story about his uncle.

> One warm summer evening, Chan grabbed me by the hand. "Come on, Kid," he said, and pulled me with him to the closet where he reached for his single-shot .22 and a box of shells. With a devilish grin he handed me a flashlight nearly the same size as me. "What are we going to do?" I asked, looking up at the six-foot-one-inch towering blond man (a chief petty officer in the Navy and a decorated hero of the first World War), who in many ways was more of a father to me than Sam had ever been. I was always a little in awe of Chan, for I never knew quite what to expect from him. Sometimes he'd scream and yell and slap me in the face, and in the very same breath

he'd hug and kiss me in an uncontrollable emotional fever. Once, he worked every weekend for five long months to save money to buy me a bicycle for my birthday. Not two days later he came home from work. I went to greet him at the door, and he kicked me all the way across the room.

"We're gonna bag us some rats."

"We're going to do what?" I winced.

"You heard me, we're going to kill some of those lousy plague-ridden rats, and I've chosen you as my back-up man." (He gave me a fatherly pat on the back as though he had just bestowed upon me a great honor.)

"Oh please, Uncle Chan, don't make me do that; I'll do anything you ask, anything! But please don't ask me to do that."

"Don't ask you to do that?" he said sarcastically. "Who do you think you are? While you're living under my roof and in my house you'll do as I say, *Do You Hear*? Now you go get a loaf of bread and get out on the porch—*Now*!!"

I was petrified and sick at my stomach, but I went to the kitchen in fear of my life. I took a deep breath and walked out to meet him.

"Now you shut up and do as I tell you, understand?"

"Yes sir."

"Tear up one slice of bread. Throw the pieces over the railing."

I did as he said. It was so quiet I could hear the rats in the alley below.

"Shhhhhhh!" he ordered. "When I say *go*, you turn on that light."

As the rats found the bread, we could see the white spots start to move. It was my job to surprise the raiders by suddenly revealing them in the brilliant flood of my high-powered flashlight. It was like throwing a spot on an escaping prisoner; they didn't stand a chance once it was turned on.

Fortunately for those rats, he was a good marksman. Usually he hit them in the head, but sometimes he'd only wound them. And while he reloaded his gun for a second shot, I'd have to hold a shaking beam on them as they dragged their broken backs along the bloody ground. I wanted to throw the light down and run screaming from the porch. But there wasn't a doubt in my mind that Uncle Chan would have picked me up and thrown me over the rail to those rats.

At other times he used a fifty-five-gallon drum filled three quarters full of water and oil, rigged with a greased tilt plank balanced on a box and perched just on the barrel's edge. At the end of the plank he placed a pile of food. As the rats walked toward the food, the weighted hinged board tipped, and they splashed into the barrel. Because the sides were slick, they couldn't climb out or hang on; so they drowned. Sometimes as I lay in bed at night, I could hear them fall, and the frustration of not being able to save them was enough to drown my own guilty conscience.

Chan saw himself as the good guy and the rats as the bad guys, and in his mind he was ridding the neighborhood of pests and disease-ridden rodents. While he was at it, he was a hunter and had a bit of sportslike diversion. In my mind the rats were simply poor old garbage eaters, not doing anyone harm. As a child I felt he was needlessly taking life, and I thought it was horrible of him to make me accessory to the fact. I was a lover of life, and as a little boy I judged my Uncle Chan as 100 percent wrong.

2

Ralph endured hostility as a boy because his family, his father in particular, did not understand his sensitivity toward animals. His father, Sam Helfer, actually found it offensive in his son. Sam had a strong definition of masculinity, and he worked for many years with fevered devotion to have his son grow in that image, characterized by women and drinking and, above all, good times hunting with "the guys." On a sobering day in autumn Sam decided Ralph was to become a man.

With gentle coaxing I was able to extract this story from my husband.

I remember more than anything wanting to make my father proud of me. As we stalked the quiet woodland I drowned in conflict. I was only ten years old, and I felt reduced to insignificance by my surroundings, dwarfed in a giant's garden. I knew with all my heart I wanted to be *with* nature, never against her. Everything here seemed so perfect, so at one with the environment. And I had been brought in as an enemy of life.

The trees were draped in long, silky strings of shimmering iridescence. Every leaf in the woods had been washed clean with the weekend rain. The logs were damp and steaming. We hiked for nearly an hour, and for that brief time my father and I enjoyed one another's company as we shared an equal passion for the woods. I hoped and prayed we wouldn't see a deer. How I prayed that day. But to my horror, as we stepped out of the rich thicket and onto the edge of a flowered meadow, there in that splendid earthly paradise stood Bambi, an enormous rack adorning his magnificent stately head.

"It's a six pointer!" my father softly cried. "A six pointer, Kid. What luck! Your first deer a six pointer!"

With unequaled grace the beautiful buck raised his head and sniffed the air. Morning mist jetted from his flanged nostrils. I wanted to be his friend, to experience life with him, to withdraw with him into his mysterious coexistence of quality and beauty; yet here I stood, weak among my fellow men, trembling in terror at inexperience, trying to justify horror by a lack of knowledge.

My father placed the rifle in my hands. I put it to the side of my hot cheek and looked through the sight at the benevolent eyes—oversized, liquid eyes that I shall always see. Caught between two worlds, wretched at the thought of killing something I loved and sick at disappointing my father, I stood immobile for a very long moment. Then, in an emotional rage I pointed the barrel toward the ground.

"Shoot, you little coward!" Sam yelled in an agonized whisper.

"No sir. I can't."

He grabbed the gun from me.

"Please don't kill him! Please! . . . Please!" I screamed. "Run, Bambi, run!" I tried to pull the rifle from my father's arms, and he knocked me down and shot. Bambi fell in a motion that kept repeating itself over and over in my mind. . . . Bambi. . . . Oh, Bambi. . . . But Bambi didn't hear me. All life had vanished from those gracious eyes.

The stinging slap of my father's cold hand against my tear-stained face shook my frozen body into ripples of pain.

"You little *creep*! Why did you just stand there? Get out of my sight! You make me sick and ashamed to have you for a son." I looked up into my father's scowling face, and he gave me a hard shove in the direction of the car. "Go on, get! And here, take the rifle back with you. You can do that much, can't you?"

The walk back seemed like a lifetime. My head spun with a sense of right and wrong. I was sure my father hated me now. I would probably pay for this day with ridicule from him for years to come. As I shuffled along, kicking roots, I was overcome with misery and a sense of personal doom.

A great white bird came out of the sky and landed on a sunny fallen tree some distance ahead. I watched him as he cleaned and fluffed his spotless feathers. Then in a moment of bravado I said to myself: "I'll show him. I'm not a coward, I'm not." I felt the air breathing heavily around me, thick and hypnotizing. My thoughts interfered with one another, and I shook with mental epilepsy and gasped uncontrollably as I aimed the gun at that snow white bird posed so generously in the drama of the moment. He was so still. And the absence of color made him stand out like a precious jewel against the green backdrop.

I pulled the trigger and shot the lovely creature. (It's amazing, isn't it, how we often become the victims of someone else's brainwashed life?) But the bird did not fall. I moved closer and shot again, and still he stood. Closer and closer, again and again. Five bullets that should have ended his life avoided him as though he wore a protective invisible shield. How could I have missed five times in a row? It must have been a malfunction of the rifle. I aimed at a pine cone and shot it into a spinning orbit. I blinked my eyes. Did the bird exist? Mirage? Illusion? I moved closer yet, then with fierce intensity focused my eyes on it. Even the violent sound had not frightened it away.

I was so close now that I could distinguish its feathers. I fired for the sixth and final time. One more bullet, one more invasion of life. One more crime against nature. And once again a miss. My bullets simply refused to kill it. I picked up a dirt clod and in frustration threw it at the bird. He flew away.

That incident was a spiritual, sacred dawning that centered upon a supernatural power far greater than anything I had ever known. If a blind man were suddenly to see or a deaf man to hear or a leaf to grow where no limb had been, perhaps he would know the shock of this illumination and how it affected me. In my altered state I began to make a speculative inquiry concerning my own human nature. Why should I feel guilty for going against my own source of knowledge? I am what I am and not what others think I should be. I was not meant to kill that deer or to hurt this bird or to inflict pain on any living thing. Today I was aware that something much bigger than I was exercising a guiding control over my life, influencing and confirming all that I wanted to do. It seemed to be pushing me with gentle wisdom in the direction of what I wanted to be. I read that event as a prophetic sign.

We sat in the uncomfortable presence of one another, and the ride home with the corpse exposed on the luggage rack of the wagon devastated me. We made a stop at a truckers' cafe, and over a hamburger Father gloated about his hunt to the girl behind the counter. I sat in empty silence, convalescing. Suddenly there was a piercing, high-pitched peeling from a truck in the parking lot as it spun onto the highway and out of sight. The smell of burning rubber was left behind. We could see the door of our wagon ajar. My father ran to the car as I looked on with a delighted grin. The rifle had been stolen.

So caught was I in the incredible chain of associations and the undeniable symbolism of the day that I knew without question I could never and would never shoot again. And so one less hunter had found his way to the forest.

The ten-year-old child's inspiration was now to become a "friend to every friendless beast," and neither his father, nor his uncle, nor anyone in the *whole wide world* was ever going to make him do something awful again.

So much of that little boy still lives in Ralph; there are times when I'm caught up in the child he used to be. But in my fantasies he is my little boy, and I give to him a perfect life and hold and hug and love my very special son.

In 1942 when Ralph was eleven, his mother left Chicago and Sam. With her two children she moved to Hollywood, the mirage of Dreamland, where they took root in the Green Apartments, just one block off Hollywood Boulevard.

Ralph always suspected the Green Apartment Building was, among other things, a house of ill repute, since the line of men coming from the Hollywood Canteen (escorted by residents) was constant, continually invading the downstairs parlor as well as wearing thin the carpet to the upstairs rooms. But his mother kept insisting that things like that just didn't happen, that Ralph's active adolescent imagination was working overtime.

That building on Wilcox and Yucca was a story all its own. Rapes, murders, and suicides occurred there. One distorted individual would periodically knock at the front door, greet Mrs. Helfer with a smile, and announce, "Hello, my name is Charlie, what's yours?" While standing stark naked, he would hand her the evening news.

"Who is it, Mother?" Ralph would call.

"Just Charlie with the newspaper, dear. Thank you so much Charles. Now you have a nice day." It must have been devastating for an enterprising young pervert to be confronted with such a complete lack of interest. But Ralph's mother closed her ears to all bad things and saw only what she wanted to see.

Carol lived on the floor just below Ralph. They were childhood sweethearts, slipping toilet-paper love letters under one another's door. In an effort to surpass reality, they'd play Tarzan and Jane in the avocado tree down the block. But Ralph never really made it as Tarzan because Carol had one of those yells that could call the wild from the nearby hills to her side. So at times, much to his disgust, there were two Tarzans in that tree. Their youth in the Green Apartments was filled with poverty and the miseries of a confined childhood. Yet they were able to console one another with dreams of a wealthy future. And when each went his own way to seek fortune and fame, she became Carol Burnett and laughed back at that launching pad.

Like a brave pioneer, without help from anyone, Ralph began to pursue a new frontier. And because he had always been ahead of his time, he had to deal with being an intellectual outcast. He was a boy led by a vision, and he had the will and the strong determination to make his dreams come true. But he was also a philosopher caught between thoughts, and his explorations took him on an early adventure of the mind.

To have set your heart on a vocation not taught in school and nonexistent as a profession is a conflicting situation, confusing in that you don't know where to start, and so you begin with a kind of handicap. Few could recognize a talent that could not be displayed. But Ralph had been set apart from other boys by providence. He believed himself to be here on a mission, and because of his determination he won the admiration of other youths around him. They felt from his strong, faith-filled nature that he was blessed in some mysterious way. He seemed in their eyes to have even escaped that victimizing domination by parents and school and the scheme of civilized things.

Many people profited by the speeches he gave them of the world as he saw it. They heard him express somewhat revolutionary thinking. These talks were the hallmark of his true personality. So convincing were his theories that people began to support them with equal fraternity and sought to find him a place to study–a place where he could practice his extraordinary ability to understand animals as no one they had ever known had.

Where could someone with this unusual natural knowledge further his education and train for the future? Where should he begin? He was a frequent visitor to the zoo and confessed often to his young companions that it always struck a depressing chord in him. It seemed to Ralph that zoos forcibly established their own necessity, for the greater the number of zoos, the greater the number of animals kidnapped and enslaved. And the greater the number of animals taken from the wild, the greater the number of zoos were needed. It represented an objectionable cycle, with man and his contradictions sitting at the head of the round table.

Apart from the zoo were pet shops, aquariums, and tropical bird stores. Ralph held down a succession of part-time jobs in some of these places and felt himself a kind of prisoner of melancholy there. A customer recommended a place called World Jungle Compound. In 1944, full of self-confidence at the age of thirteen, Ralph began weekend bus rides to Thousand Oaks.

His countenance began to change with the brightening possibility that by being in such a place his ideas could begin to develop. Being around animal people, he felt sure, would bring about stimulating conversations and the comfort of a common brotherhood. In his youthful mind he naturally surmised that anyone associated with animals must think kindly of people, too. He began to visualize a great transformation in his future.

4

World Jungle Compound had been known to the people of Los Angeles since 1926 as Goebels' Lion Farm. Louis Goebels had bought out the old Universal Zoo when it closed in 1925 and the Ernst Lion Farm that same year, combined the two, and launched a famous animal career.

In 1950 he sold Goebels' Lion Farm to I. S. Horn, who allowed people to believe him to be the son of the famous Trader Horn. In fact, he was not related, but he capitalized on publicity from a very successful film. With the farm's purchase the name was changed to World Jungle Compound. Louie retained the enormous barn at 2382 Pleasant Way (which still stands today) and concentrated his efforts on one of the largest import-export businesses in the country.

Frank Buck (bring 'em back alive) and Clyde Beatty were the great heroes of those days and close friends of Louie Goebels. Ralph used to hang around that barn for endless hours in hopes of catching a glimpse of them, and when they came in either to drop off or pick up an animal, he went out of his way to help, in a faint hope that they might take some notice of him.

Clyde Beatty was a short man, whose complex over that fact was apparently appeased by the work he did in the arena. He seemed content and gratified. He began to take Ralph under his wing, and as one would try to set right and heal the misinformed, so he would venture to show Ralph the way to train animals. Although Ralph was just a kid, he and the famous Clyde Beatty would on occasion enter into conversation.

But Ralph was against all that Beatty preached regarding the arena relationship. He was of the belief that animals could perform because they wanted to and not because they were made to. Even though this outrageous opinion set Beatty's teeth on edge, Beatty still attempted in a kind way to convert Ralph to the only way then known of handling wild animals. Since he was the great Clyde Beatty, who was Ralph to question his methods? Through lack of experience, Ralph almost relented, but never quite. Only years and the knowledge to come could prove him right. He waited years to say, "Clyde, I know a better way." But Clyde had died when things were clear to Ralph. I don't think Clyde would have changed his mind anyway. I'm sure he would have said, "Maybe it works for you, Kid, but that's not the way for me."

In the days when Frank Buck's trucks pulled into the Goebel barn no one was concerned with the fact that the zoos, circuses, and natural history museums were depleting the globe of her wild splendor. No tears were shed, no weeping.

Caged and lying submissively on piles of defecation, surrounded by stench, they arrived–hundreds of thousands of them. From here they were distributed–that is, all those who survived the tedious ride across the sea. The market and the demand were great, and the dealers filled every order with broken bits of life.

In and around 1946 Horn sold his shares to a group from Twentieth Century Fox, and the World Jungle Compound became Jungleland. Excitement glamorized the town as big circus traveling trunks came to rest at Jungleland. Anyone who *was* anyone in the circus world winter-quartered on these grounds and trained for the new season.

Every spare moment of Ralph's life for the next few years was donated to cleaning cages, sweeping walkways, emptying trash barrels, raking pens, and working in any menial capacity simply so that he could be around animals. Here he was able to see how training was done but he felt that something was wrong, and not knowing any better, he began to do it the way everyone else did, nearly getting himself killed at the very start.

Although each trainer loved to boast of having developed his own method, the principles of arena-breaking were nonetheless the same. Two- to three-year-old wild animals "fresh caught" and brought directly from Asia or Africa were preferred. By the time the cat had reached the States, he was quite used to the confinement of a small cage, having traveled the ocean in a crate no longer than himself. When the lions, tigers, and leopards arrived, they were put directly into a squeeze cage (usually a 3' x 6' barred enclosure with one moveable side that cranked inward to squeeze the animal tightly against the other side, making it impossible for him to move.) When the occupant was adequately smashed in, a reinforced collar made from three-inch wide double-stitched leather was buckled about its neck, a steel ring having been slipped through the collar first. A four-foot piece of chain trailing a long rope was then attached to a ring that had been bolted into the arena floor. The rope was then threaded through the heavy bars, where it was securely fastened outside. When the cat was released from the squeeze cage through the chute and into the arena, the slack in the rope was pulled up, and he was left with about a ten-foot extension from the center ring. There were a pathetic few minutes of hysterical reaction from the feline, who flipped about, thrashing to free himself from the restriction. Exhausted from the useless effort, he crouched closely to the floor and flicked terror-stricken eyes at his surroundings, hating everything and everyone he saw, snarling or barking at every unfamiliar move.

It was then that the "fearless" trainer would enter, equipped with a five-foot hickory stick, a chair, a gun (holstered), and of course the ever-popular whip. The cat, who by now despised man, would lunge at the trainer in an all-out effort to kill him, the frustrating chain always jerking him back. The trainer positioned himself at the end of the animal's reach, and with each lunge, he beat the cat with either the hickory stick or the whip. Having sufficiently reinforced his masculinity, the man would exit, wait half an hour, enter, and repeat the cruel process over and over again. (Watch me! Watch me! Chest expanding, ego inflating.)

This barbaric lesson ended only when the trainer entered the cage and the cat, rather than attack, backed away in submission. Instead of *kill man,* it was now *fear man* or be beaten.

From that point of retreat and after one more treatment in the squeeze cage, the cat was set free from his shackles. Next he was released from the tunnel to the brutal arena, where he was observed to see which side of the cage he favored, the right or the left.

On the day the seat-breaking program was to begin, the trainer, having determined his pattern, would place pedestals accordingly. The frightened animal usually panicked when he first saw the structures, which represented the unknown. Distrusting as he had been taught, he would either hide beneath it or knock it over. The process took some time for the cat to understand, but once the animal took his seat and the trainer backed away from hazing, it wasn't long before the cat realized that the pedestal was the only safe place in the arena, and he was allowed to relax there. From that point on the training was easy.

So fear by reprimand was established first. *Respect man or he'll hurt you.* This is what Ralph refers to as *fear training,* and this is the way many cats were broken—through animosity. This was also the way men were killed. The animal hated men so, he would sometimes wait for years until he saw the moment the trainer's guard was down—a back turned, a whip dropped, a moment's hesitation. Then *revenge*! It was a savage distortion before the age of enlightenment.

Ralph was training in veterinary medicine when he met Mabel Stark. Her tragic image hangs heavy in grey shadows still. She must have been at least fifty-five years old in those late '40s, for I remember having been told that in and around 1911 she had stopped a successful nursing career to join the circus. Ralph said she had a special knack for screaming four-letter words and making everyone feel like an incompetent idiot, but he thought she was absolutely wonderful. She was the only woman tiger trainer in those days as well as the only trainer who had actual touch contact with her cats. She never carried a whip, gun, or chair. Her only means of protection were two small training crops. Both Mabel's breasts had been removed as a result of severe maulings. But in spite of having been attacked many times, she never shied from the arena. It held for her a strange kind of ecstasy, nothing to compare with other desires.

I met Mabel Stark fifteen years later. It was after weeks of discrimination, and after I had been able to convince Ralph that I should become a member of the work team at the ranch. Through subterfuge, I had been appointed as head of the Animal Care and Keeping Department.

After what seemed like an endless time, I became aware of what was happening. I had been given this job to discourage me. Apparently there were those who looked upon this assignment as no less than latrine duty. It was true that this position involved all of the unglamorous aspects of life at the compound, but I have always loved playing in mud; so the job and I were in complete accord.

Before the days of packaged animal food products (which Ralph helped to develop) I had to drive our truck, old Bessie, sixty miles to Thousand Oaks and to Jungleland where we bought our meat for the cats and canines. Jungleland was by then an empty place, filled with the ancients of the circus world. They came there to die.

About the park were cheap, scratchy recorded jungle drums and false animal sounds. Antiquated barred cages lined the entrance walk. Their occupants lived with mange and in filth. There were more public complaints about Jungleland than there were customers at the gate. Outside, just next to the box office, there was enshrined behind glass a colorful poster revealing a pretty young girl training a dozen tigers. I remember that large, black, foreboding arena that ate most of the performers who went through its doors.

While Benny, the blood-soaked butcher (who always looked like a predator after the kill, for he was covered from head to toe in gore), loaded Bessie with red flesh, I went once to observe the show. I sat reluctantly upon a splintered, stained, dirty bench, waiting to see the famous Mabel Stark, whom I had come to know through Ralph's tales. A cage boy slithered out, his hair so oily it dripped down his forehead; his pock-marked face was outdone in ugliness by lidless, beady eyes. He appeared vile and nasty, like someone you never wanted to know. He dashed about with a long pole in his hand, and as he pulled the chute the release the tigers, he jabbed each one and sent it through in a rage. There was a loud circus fanfare and an offstage announcement: "Ladies and Gentlemen, the world's only woman tiger trainer, the remarkable Mabel Stark!"

The applause, not exceeding four claps, faded to a low murmur as an old, weatherbeaten lady wearing jodhpurs, a long-sleeved white shirt, and knee-high riding boots entered. A dull crown of very bleached curly hair was pressed tightly to her head. Her powdered face was wrinkled and terribly worn. She stepped forward after pedestaling the cats and bowed low. I found the performance, while certainly admirable for someone her age, quite sad, and I suddenly thought, "There, but for the grace of God, go I." But without a doubt the worst thing about Mabel was meeting her. After I had spoken with her only a few times, my overwhelming impression was that she was vulgar, crude, coarse, and hard, to name a few of her good points.

I have long since learned not to take affront at such a generalized reaction, since for the most part the old animal people are a breed all their own. And after all, it was not a personal confrontation; it was simply that I was not a tiger.

When in the mid '60s Mabel was relieved of her employment because of her age (at the insistence of the park's insurance company that if she were to remain, they would cancel their coverage), she took it as a terrible blow. She was completely annihilated, for all the tigers belonged to the park and not to her. She was barred from any contact with them other than as an observer from the outside. The old circus dowager had practically put Jungleland on the map, and everyone felt her release was an unkind, emotionally brutal act. One horrible incident led to another. Shortly thereafter her favorite tiger got loose (within the closed park), and rather than call her to capture him, the park attendants shot and killed the cat.

Mabel Stark put a plastic bag over her unhappy head and committed suicide. It wasn't long after her death that Jungleland died. And a great distance of time has elapsed since I shopped at Benny's slaughterhouse.

5

Before Ralph made his entrance, the animal business in general had been looked down upon in the United States. The principal examples of animal people at that time were, for the most part, a group low in the estimation of most people. Some of them operated dingy roadside zoos, and many displayed as much regard for their animals as they did for cleanliness. Many were missing fingers, an eye, or limbs. A majority of them reeked of alcohol, chewed tobacco, wore filthy clothes, and were addressed as Slim, Curlie, or Ace.

In and around 1945 when Ralph was only fourteen and volunteering his time to anyone who could get him on a production set, there lived in town two middle-aged brothers, Dave and Curley Twifer, both of whom were realtors. They were "backyard pet owners" and pursued an interest, primarily for pleasure, in exotic animals. Among their collection were several large baboons, one of which attacked and killed Dave, leaving Curley (who had an aberration of epilepsy) alone to work an occasional movie job. His malfunction, characterized by recurring attacks, could happen at any time. Ralph was on the set when Curley brought in a mountain lion (that only Curley could handle). And as he stood talking to some friends, the lion by his side, he fell upon the ground and began to convulse with violent muscular contractions. Curley fell just next to the cougar, who, upon seeing such a sight, left at a high-moving speed that increased to leaps and bounds as he demolished part of the set and scattered its crew in all directions, retreating from the spasmodic Curley.

I was hypnotized by the activity taking place before me. Someone had propped open Curley's mouth with a stick and was at the same time holding his tongue as two others held his legs and one man pinned his shoulders from behind. I couldn't help wonder how anyone with such an affliction could be in the animal business.

The director came running over, and since I was the only one standing, he grabbed me by the shoulder, diverting his attention for one brief moment to Curley and–apparently finding the whole thing too bizarre to even comment upon–issued orders for me to capture the cat–who at this point was heading for the fifty horses making a Western next door.

"Me go get him? What do you mean? No one can handle him but Curley." I volunteered to switch places with one of the guys restraining Curley. No chance!

I picked up my leaded feet and tracked the mountain lion. Everyone was yelling incoherent directions at me. "He went here, he went there." Here cougar, there cougar, nowhere a cougar to be found. One man swore he saw him run into the old livery stable on the set next door, the doors of which I burst through and promptly fell over the cat, who lay just on the other side. Scared him, scared me! He split to the back of the barn, not even so much as looking at the horses, which were panicking about with wild-eyed apprehension, and the old barn began to come apart as they back-kicked and reared in their stalls. I began to yell with abandoned flair, screaming in falsetto, "He's in here! Quick, someone, he's in here."

A less romantic figure I could not have expected to appear,

as Curley crashed through the barn door, knocking me far to the side and into a pile of hay. The stick from his mouth was now in his pocket. "Thanks kid. Ya done good," he said to me as he and his mountain lion walked back to the set.

Not one month after that incident another film called for an eagle to fly across a lake from the top of a tree, and who walks onto the set with a golden eagle perched upon his gloved arm? None other than Curley. I couldn't believe it. I thought surely after that incident no one would hire him again, news spreading the way it does in this town. Nevertheless, there was Curley, climbing to the top of a rather tall tree. The director had a megaphone across the lake, and everyone stepped out of the shot as he called in gutteral intonation, "Are you ready?"

Curley replied, "Yes, I'm ready."

"Action! Okay, Curley, let him go. . . . Let him go, Curley. . . . Curley?

We squinted our eyes and focused on the tree across the lake where Curley and the eagle should be sitting. It was shaking to such a furious rhythm that the leaves were deserting with each epileptic jolt.

We ran toward the trembling foliage, and there, wedged in the fork of two branches, was old Curley passing through one of his fits. The eagle sat one flight above, looking down at the shaking vigorousness of the attack that was just about through below. As Curley came out of the vibrations, he climbed back up, put the bird on his arm, and yelled, "All ready over her!" As though nothing at all had occurred, he flew that bird across the lake, where it landed on a preset perch.—That's show biz.

Ralph said Willard the Indian was one of the meanest men he ever knew. He wore khakis and a pith helmet, never smiled, had an iron jaw, was as strong as he could be, and carried a thick, heavy stick with him wherever he went. Willard was a chimp trainer, and whenever he worked on a set, the director would explain the action: "I want the chimp to start from here and walk to there," which was certainly simple enough; and although Willard was simple, nothing was simple to Willard. When the director called "Action," Willard would raise his stick and zonk the 140-pound chimp on the head, at which point the chimp would walk across the scene like a confused alcoholic. Actually, that ape was quite decent about those cracks on the bean. In fact, Willard said he wouldn't work without them. It rather reminded me of a shooting gallery, where the little target walks in front of you and every time you hit it, you ring a bell and *boing,* it turns around and ambles the other way–*boing, boing, boing.* But the day the Indian conked an orangutan in the same fashion, the orang did not find it appealing, and he spun beneath the blow, snatching the stick from Willard. Not only did he beat the meanness out of Willard; he nearly separated his limbs from his body. It was Willard who staggered around the set that day.

Many of these people took enormous pride in elaborating on the time "that there bear tore this here finger from my hand." And on the set if they didn't receive some kind of injury displaying such bravery, they insured themselves by carrying a razor blade in their wallet; when the attention of the crew was diverted, a quick slice of their arm added to the flavor of the day and increased their studio fee. "Why, he mauled me, you seen it, lookie here–blood! That must be worth sumpin'."

The handful of outside people with whom I was to come in brief contact were in general cold and hard with one another. Vulnerable, they tried to compensate for feelings of inferiority. Each distilled upon the industry his complete ignorance of his craft.

There was little contact among any of them. They did not swap stories or exchange information or even talk to others who supposedly stood on common ground.

And there were such distortions then. One woman accepted donations of unwanted animals under the guise of a nonprofit organization. The stipulation in the "adoption papers" was that the donors were to supply a monthly fund to support the animals, and there were "no visitation rights." She put the majority of the animals to sleep and lived off the free flow of blood money.

One organization turned another of the same ilk into the powers that be, unjustly accusing them of "cruelty," making up an ugly story that later proved to be a lie. Through this perversion they attempted to eliminate their competition. The shallow-minded public demanded an investigation and devoured the innocent victim. A sudden strength was thus given the traitor; his false words were printed. He could speak now from a perfidious pulpit, and so he thrived for awhile, as parasites do.

There were organizations masquerading as champions of all animals. They were usually self-indulgent pragmatists, praising their own deeds and disqualifying everyone else, often self-imposed martyrs hiding behind the rape of the wild. Sentimentality and ignorance clouded the public's vision.

5

Having someone in your family encouraging you when everyone else is against you and trying to make you into something you're not is very satisfying. Uncle Erv believed in Ralph, and he watched with increasing interest through the years as Ralph gathered knowledge through working at pet shops and in Jungleland. Now middle-aged and never having achieved those dreams of his own, he invested in his nephew; together they opened Nature's Haven Pet Shop at 151 North Western Avenue. It was the only pet shop in town that sold exotic birds and tropical fish, reptiles, assorted monkeys, and an occasional kinkajou and raccoon.

Nature's Haven signified a contagious moment in Ralph's esthetic life; it was his first opportunity to express himself. Those years working for others, in disagreement with them, had been a bridge to his own place. When those first animals arrived, he was beside himself with joy. But Ralph soon found himself becoming so attached to the animals that he couldn't stand to part with them. It was not at all what he had thought it was going to be. When a customer came to purchase something, he scrutinized the person's ability to care for the animal to such an extent that more often than not he killed the sale.

Several years passed by. Ralph tried with growing frustration to make a go of the shop (if only to please his uncle). But his dislike of selling the animals and the routine of a restricted store life were making him miserable. Working in the city was another deprivation, for his wanderlust was beginning to get the better of him. He felt responsible for having involved his uncle in such a venture, one that was driving him to distraction.

Then came the excitement of those first studio calls, from a movie industry so desperately in need of animal actors that they phoned a pet shop. With the little experience he had gained (up until then he hadn't realized just how little that was) Ralph began to embark upon his career. The Industry was a way in which he could see making money doing what he loved best, being with his own animals–animals that he would never have to sell again.

"When the call came in to work a lion for a feature film," Ralph said, "I absolutely jumped at the opportunity. Although I had a veritable zoo in my pet shop, I certainly didn't tell them I had no lions."

> As I searched for a lion, someone told me of a man who owned a circus cat and kept him in a garbage dump just

outside the valley. I found him in a deplorable condition, living in an infectious pile of trash. I rescued him and began to work him.

Everything was going well on the set that day. The scene called for a carnival setting in which I was to double for the star. The lion was to be hazed up onto a pedestal in the Beatty-Stark fashion. That seemed easy enough. Cameras were rolling, and the cat performed beautifully. I mocked Beatty with the whip and chair as I'd seen him do so many times. The director yelled, "Cut!"

"That was a piece of cake," I thought to myself. There was nothing to it.

It all had a romantic and heroic air about it. I strutted, as all great trainers do, you know, until someone shouted, "Look out!"

As I turned in the direction of the voice, the full weight of the lion hit me from behind, knocking the breath from me. Before I knew what was happening, I saw blood gush from several spots on my arm. I didn't feel anything and was about to get up when the lion, who had backed away, flew at me again.

This time I saw his dagger fangs go straight into my shoulder as his claws clung to my side. People were screaming and fire extinguishers were blasting until I saw only a thick white column of smoke rising around and above me. Blood and saliva, hot breath, and terrible growls engulfed me as I became entangled in a dense mat of fur that ripped and tore at me as we thrashed about. There was no time to think as I twisted in anguish and fought desperately for my life.

Haze began to cloud my vision. Stages of dizziness overcame me. I began to feel very weak. Then I was beyond pain and suffering. I simply wished it to end. I actually wanted to die.

On three separate occasions I came to, in the hospital, having dreamed I had again been attacked, mauled, and badly bitten. And each time I took the incident personally. Why? Why? Why was this happening to me? To me, who loved animals more than anything. Each time I stood up, one knocked me down. Why were they trying to kill me?

Then quite suddenly from my near deathbed came the answer: "Helfer, with all your theories, you did it the way everyone else did–arena style." What I had done was against all I had ever preached.

6

San Fernando Valley awoke one morning to sounds from another continent. Ralph and Uncle Erv had dissolved their partnership, and Ralph and his animals were embarking upon the adventure he had been pursuing his whole life. This is his recollection of the beginning of his new life.

Only two acres; 15924 Armenta was small, but it was mine. And the old dog kennels with "runs" sat between the house that had no furniture and the pasture with no horse. Orange blossoms perfumed a smogless air then. I bought an old Plymouth station wagon and painted *Nature's Haven, Wild Animal Rentals* on each side. My car, my ranch, and I were now in business. Each morning I couldn't wait to start the day, for now everything had substance. And from that day on I was forever racing against time.

I began, with fractured limbs, to scrutinize what had gone wrong. Training was definitely not a spectator sport, and if I was to become an active participant, I should have to match energy with instinct. The first thing I decided to do was give away the wild stock I presently had. By *wild* I mean animals such as the lion that had mauled me and several other unmanageable ruffians I had quarreled with. I gave these dirty half dozen to a zoo without regret. I could not be an emotional, fanatical animal lover, feared by none and rejected by all. I must be in control, never mixing fantasy with reality, or I would become my own victim. I began anew by raising cubs and establishing myself in a position of leadership like that in a well-adjusted parenthood. I believed attitude to be an important defense. Having seen for years the kind of training I completely disagreed with, and having suffered the final humiliation under attack, I was convinced that using force was wrong. It merely fed the ego and resulted in indifference to the animals' feelings and a preoccupation with aggression—a predator's rank!

My object, then, with gentle breaking, would be to work the animals on the often storm-filled set without duress and with the confidence that what I was doing was safe for all concerned. My methods included love, patience, and understanding based on mutual respect. As I grew with those first few animals, I found training with affection rather than fear to be most gratifying. I was convinced now that man's ego had created arena training, and the result was always the destruction of an animal's pride. The trainer was the only "winner." After all, hadn't he conquered the will of the so-called "ferocious beast"? If the animal didn't show ferocity to the blood-thirsty public, what, then, was the challenge of the arena? And the flair on the performer's part, with whips that popped excitement and guns that added explosive thrills—anything less grand would surely find the audience asleep. They wanted thrills and chills and the grisly possibility of seeing someone torn apart before their eyes. Nothing had changed since the days of the Coliseum. The public demanded ferocity, and so it was given to them on a grand scale. Thumbs went down in a succession of boring yawns at anything else. I knew "the only beast in the arena was the crowd."

And so I was to discover that fear training and the ways of the pedestal were not the only ways. I found the reasonable assurance of the gentle, hand-raised animals, who, in their relaxed state, had their wits about them in a controlled situation, where one must adjust to the unexpected, not promote it.

It was then that I developed "affection training," wherein my main weapon became the use of vocal inflection, using high-pitched sounds when a job was well done and a deep, scolding voice for a bad attitude. I never used a physical reprimand, where pain represented the challenge and defeated the purpose. The press coined the words *affection training* because it sold to a commercial world. *Emotional training* describes better what we do, since we deal directly with animals' and peoples' emotions. Nine times out of ten animals perform because they want to and not because they have to—the tenth time being that of the stubborn attitude of a spoiled child, who, if allowed to get away with it, glories in "one-upmanship."

In those days a crew of dirty handlers loaded steel-barred cages with heavy cranes onto flat-bed trucks and took the "wild animals" to the sets. Jungles were built within arenas, and actors were attacked by stuffed animals, thrown at them from above. Trainers worked with young cats that could be backed down and frustrated, and the split-screen process was having a field day.

The animal work on the set then became not only costly but extremely dangerous. The business had exhausted itself. Those trainers had gone as far as they could go. It was time for a change, but an individual voice in a world of community opinion was not a voice to be heard, unless that voice could back theory with experience. So Ralph decided, "I'll show them all."

Times were rough then. Ralph was ridiculed by other trainers, who were livid that he had ventured so radically from the mainstream. They spread the word in wildfire fashion throughout the studios about a young, punk kid who had not only invaded their territory but was trying to change their world. He had the unmitigated gall to question their techniques. If a studio had any concern about production, officials would have to be mad or desperate to call Ralph Helfer.

They did not know they had been caught up in a slow process of deterioration, which led to bankruptcy. As a result, in 1951 Ralph took over the bulk of the wild animal work in Hollywood, and went ahead developing his theory as though the whole universe supported his ideas. The challenge was shaping Ralph's fate by a strange means. With all Jungleland's handicaps, she held a precious period of unequaled observation. It was a time of visual acceptance and of emotional, mental rejection. But isn't that stimulation of disagreement a part of truth? Having now become absorbed in the ways of captive animal training, Ralph began to develop his own theories. And it was fortunate that he had the comparison to draw from. Others were expecting man-motivated response, when in fact they would have progressed much further by penetrating into the animals' world instead of catering to their own. Ralph felt that he could establish an orderly hierarchy, based on mutual respect instead of fear. The new system would, he believed, lessen disputes among animals and, hopefully, control their anger to a certain extent through a pecking order close to that of the animals' own.

Filled with questions, he was led by inner forces to study animals in the wild, adapting their social ways to his captive, controlled environment.

> I began by entering into their world, centralizing myself there for the moment and excluding my power of choice. If I was to succeed, I felt I must know as much regarding their world as I could possibly comprehend. And being man and having long ago shared the wild with them, I would abandon

my man-ways in the hope of understanding more of theirs.

The order and disposition of nature is such that when I became totally submerged in it I felt I could not go wrong. Like the other animals, I was not in doubt. Things were as they should be. We ate as instinct told us. We followed the trails we were meant to take. We guarded only our own territory and seldom infringed upon others. We didn't consciously seek our own destruction. We had a pure, untainted, golden extension that linked us directly with the real power, and we shadowed in that order and kept in harmony and peace with all things. I concluded that the animals were programmed for life as was the oak tree and that life was enhanced by reality. In my affair with nature, I strove to cure myself, not in an act of vengeance, but through an act of God.

I found that to judge an entire species by one animal would be like judging the human race by the behavior of one man. Although the potential is there in all, the development is completely different in each. Above all else we must recognize the status of the individual and never compare his advancement with others like him.

I developed a detailed daily schedule to fit each animal, based on a reinforcing support system and the growing close association of our relationships. I attempted to create happy, relaxed moments by reinforcing the animals' trust in me, stabilizing my intentions by a touch and by a passive vocal inflection. In this early learning process I wanted them above all to enjoy the experience. The lessons were short, never emotionally exhausting. Too much of a good thing could result in overtraining and defeat the purpose.

One of the great advantages I had over human psychologists was that I did not have the task of relieving guilt in my students. Fortunately they did not carry that weight. The range of behavioral disorders in animals is not broad as in man's sophisticated state. In animals the basics are for the most part a simple alteration. If you care about and know your subject, have patience, and establish a mutual respect, you become a growth center. Contrary to the human psychologist, when your pupil is experiencing good mental health, you cannot wish him well and send him back into the world. For once you have conditioned an animal to your social environment, you are responsible for that animal until the end of its days.

Ralph's animals have great affection for him.

I brought up my animals as one might wish to raise his children. We built a strong foundation between us and knew one another as friends. I discovered their individual idiosyncrasies, their likes and dislikes; and I developed ways to relax their fears.

They rode in my car with me and were introduced to blinking lights, honking horns, unfamiliar odors, unusual sounds, and all the unnatural aspects of society—so that when they finally walked out into it, on a leash as one might take his dog, they walked out secure with me, their friend, and they reacted as well-raised little ladies and gentlemen, responding with gusto (as a child who wishes to please his parent) to the verbal and hand commands: "Sit," "Come," and "Stay."

I wanted to create something that would satisfy me and yet would blend with my vocation, rather like an artist who not only is able to paint but can also teach, or a vocalist who can accompany himself with a guitar. I wanted to put a song filled with meaning to my work.

Although every day has brought continued confirmation that has advanced the science of affection training, certain incidents in the early evolution of my theory still scare the tar out of me. One of these concerned Tammy, a magnificent 400-pound African lioness, perhaps the most exquisite of her kind, who would have been, under other circumstances, a huntress of great and noble splendor. Her senses were keen, and her nature was gentle. She was four years old in that early haze of a Los Angeles dawning when we drove at speeds in excess of fifty-five miles per hour. We were on our way to the studio via the San Diego Freeway. Tammy lay comfortably vibrating on a bed of straw in the horse trailer my station wagon was pulling. Unexpectedly, and for no apparent reason, the trailer suddenly came loose from the car hitch. But the chain had remained attached, and I swerved and veered and swung about, trying to avoid a jackknife.

Thinking I had everything under control, I hit the brakes, and as I did, the trailer toppled over and settled on its side. The back door popped open, and out burst one very startled lioness, straight onto the busy San Diego Freeway. Drivers lazily on their way to work, coffee in hand and eyes half-masked, were shocked into a sober awakening. Brakes screeched, motorists pulled to the side of the road, horns honked, ladies in rollers screamed from inside closed steamed windows, and the putrid odor of burning rubber filled the air.

Tammy stood bewildered in the middle of the four lanes of backed-up traffic. I literally flew from my car, calling her name. When she saw me, she took one last look around at the unbelievable problems of civilization, and with behavior unbefitting the queen of the beasts, she leaped and bounded to my side, grumbling and groaning all the way, as if to say, "Did you see that?" Four hundred pounds of lioness sat trembling next to a 150-pound shaking man. She couldn't get close enough to me. If I had been made with a zipper, she would have crawled inside. I wrapped the belt from my pants around her neck as a leash and led her to the open station wagon door, where she jumped into the back–and we were off.

Not to mislead anyone, there have without doubt been those who long before me were handling animals with gentleness and a mental awareness between them, but it was within a narrow frame and certainly had never been employed on a large scale–

had never been utilized outside the arena or beyond the backyard. Above all, third parties had never been brought into it. Only a one-to-one relationship had existed between man and animal. Until now wild animals had never gone relaxed into the social reservations of man, the madness of an intricate city, the chaos of a hectic, fast-paced studio existence–into planes and trains and other moving things. And because they had been introduced to life in a multifaceted way and had been conditioned to it as we have, they went confident into the relocation of man-made nature.

So people began to say, "You can do it with a lion, perhaps, but you'll never condition a tiger, or a leopard, or a bear, or–!" So a good part of my inspiration was to prove them wrong, and my efforts eventually led to the release of stress among many exotic animals in captivity.

I have never intended to imply that captivity is anything more than second best. But until greater numbers of animal lovers with a social conscience exist on this globe of predominantly wild-animal haters, captivity is possibly the only protection for wild animals. But unless there is a reason for them to live in captivity, man will eliminate them entirely. Zoos allow areas for display and research. We have developed a place where man-animal coexistence can be observed, where behavioral studies can be carried forward as never before. Hollywood, so often considered a stepchild, has encouraged and wholeheartedly supported my development. Although scientists may look down their well-schooled noses at such a training ground, thanks to my town, exotic animals have one more reason to live.

7

At Armenta Ralph's animal corporation and his philosophy began to resolve the movie industry's dilemma. Many of his animals came to him through unusual circumstances, and a good portion of them were donated.

Toby and Jenny were two wonderful old burros who had been together all their lives. One never went anywhere without the other. Ralph had found them attached to a tremendous cement wheel, just below Hanson Dam in the San Fernando Valley. For ten hard years they had been walking in a circle, tugging and straining beneath a hot sun to pull a weight far greater than their own to grind a Basque farmer's corn. Toby already had an overdeveloped muscle that had partially collapsed his crest, and Jenny's back was beginning to sway. They smelled of the sweat of the working animal, and their skin was raw in spots from the years of rubbing where they were strapped into their heavy harnesses. When they were freed from bondage and turned loose in the green field between Ralph's house and the kennels, they jumped into the air, kicked out their feet, snorted and hee-hawed, and carried on happily from that day on.

It happened that a taxidermist had been stuffing two lions in his shop. A couple who had recently returned from a hunting safari in Africa had sent the bodies of the lions ahead, preserved in a refrigerated crate. As payment for mounting the animals in lifelike form, the taxidermist was given, among other things, a little male cub, a child of the two they had killed, which they had brought back alive. The taxidermist traded the orphaned lion to Ralph for a case of wine.

With the entrance of that animal into Ralph's life, there began to grow great emotional ties. The little lion was called *Zamba,* after the Zambezi River, beside which his noble parents had died.

Ralph's first horse was called Sahara. He had wanted a horse of his own for as long as he could remember, and now he finally had one. She came to him from the stables in Griffith Park where, through no deliberate act of her own, she had nearly killed a little boy who was too inexperienced to handle her. When he turned her to head for home, he fell from the saddle and, catching his foot in the stirrup, was dragged into unconsciousness. In the history of the animal business, man has usually blamed the animal for accidents involving them with people. Man seems to believe the animal–in his primitive state!–causes the accident intentionally, and so that he will never do it again, he is sentenced to die, the victim of an ignorant court. The stables reported to the child's insistent parents that the "killer" had been put to sleep. He had, in fact, been given to Ralph by the sympathetic horse owner. Sahara was an exquisite Arabian mare. At the crack of nearly every dawn the young man and his horse raced through the open field next to the train that steamed through the valley where freeways crawl today.

The humane department also gave Ralph exotic animals they had confiscated, as did confused private parties who had suffered bad experiences with their wild pets. (Let me interject that we do not now, nor have we ever, believed in the "wild animal pet." What we do is a profession and should be treated as such.)

Many actors owned unusual animals in those days. Mae West and her sister gave Ralph his first chimp, Coffee, and Debbie Paget gave him the chimp Lord Murphy as well as Haji Baba, a mischievous Golden Gibbon.

When Ralph began work in television on shows such as Red Skelton's, Jack Benny's, Al Jarvis's, Billie Burke's, *You Asked for It, Magazine of the Week,* and countless others, he worked many times only for the opportunity to express himself. At that time the phone rang off the hook. People called who couldn't handle the cute little lion who had unexpectedly grown before their very eyes and was now terrorizing their life. Hollywood began weekend visits to Armenta to play with the animals and pitch in with the chores. Young Elvis Presley was a regular visitor, as were John Agar, John Russell, Sabu, and dozens more. Ralph and his animals held a fascination for many people, who came to learn as the curious do facts of importance, many of which had a lasting influence on them.

Armenta became the paradise of Ralph's childhood and the experiment of his adolescent youth. Legendary animals and an extraordinary, fantastic way of life began to fuse themselves, and the faith to pursue his beliefs slowly became an undeniable reality. Two years of wonderful experiences happened there (too many to attempt to record here). The San Fernando Valley began to grow, and as it grew Ralph's permits became restricted. Wild animals in a residential area were zoned against, and understandably so. At that point he had a far greater number of animals than the little compound could support anyway. He searched for a place to expand and found Brookins.

By the time Ralph left Armenta, he took 100 not-so-wild animals with him. And at the same time, another misunderstood species came along. That cluster of lonely people became a group of devoted disciples.

Dozens of old estates dot the Malibu Hills, but the 400-acre Brookins, at the top of Mulholland Ridge, was by far the most spectacular.

The ancient, weather-tossed, twenty-room Spanish hacienda had been built in the traditional shape of a horseshoe. It hovered 'way up on the hill just at the property's edge. Pepper trees rocked in the courtyard, their switchlike branches often lashing out in windswept anger. In the valley far below a five-acre lake filled with deep waters reflected everything above it and beckoned with refreshing temptation to everyone around. In the spring of '54 Ralph placed his animals between the lake and the house. At night their wild song echoed through the Malibu Canyons.

All those enthusiastic people who regarded Armenta a retreat brought recruits to the nest above the sea, to experience together the serene isolation of his paradise, which lay close to the sun, veiled in early morning mist. Soon the twenty rooms were filled to capacity with those in search of nature and her solitude.

Brookins seemed the ideal place for Ralph and his animals to live. He suffered then from delusions of grandeur.

Ralph was regarded by some of the Malibu folk as a bit of a natural phenomenon and became the target for endless questioning. And since he held a weekend job at Trancas Beach, working in a hot-dog stand to help make ends meet, many of the hill people snacked there to catch a glimpse of the young man whose legend was beginning to grow. ("Hey, Floyd, that's the weirdo who lives on the ridge with lions and tigers and bears. Jeeze, I'd sure like to see what's going on up there.")

Some of the cattle and sheep farmers of the area would call and trade their dead animals for a visit to the place. Once while Ralph was pitching hamburgers, a man stopped by to offer two goats who had been accidentally struck by a Sunday driver. If Ralph wanted them to feed to the cats, he'd have to stop by the man's farm and butcher them out. And since Ralph hadn't driven Old Bessie to work that day, he could leave the meat in the farmer's cooler until he brought the truck down. "But ya gotta butcher 'em out tonight, Son, or they'll bloat, and there ain't no sense in lettin' good meat go bad."

"Yes sir, I'll stop by on my way out and thanks very much for thinking of me."

"Think nothing of it; it's a pleasure to do a neighbor a good turn. By the way, would it be all right if I brung the missus and the kids up to yer place for a visit sometime?"

"Why, sure it would," Ralph replied.

Ralph's friend Jamie was enjoying a day of leisure at the beach and was meeting Ralph at 6:00 p.m. to hitch a ride back to the ranch. Jamie was the product of an unhappy home. His father was a strict intellectual who had little time for his son. His mother apparently suffered from the same lack of love; she seemed always involved in extramarital activity. It wasn't so much that Jamie was a problem child as that he had grown up in a problem atmosphere. He struck back at his family with defiance and rebellion. Both parents were so caught up in their own selfishness they weren't able to cope with Jamie. They agreed to send him to a delinquent boys' camp for their sanity.

"I have a better idea," Ralph said. "Why don't you let Jamie spend some time on the ranch with me? I think the animals, the activity, and the unusual pace of my life may help bring him around." They gave the boy over to Ralph with their blessings and some unemotional forewarnings.

Jamie was now in his second year at Nature's Haven. After having spent only a few weeks at the compound, he had been known to shout out loud things like, "I mean this is what's called being alive!"

That evening Ralph borrowed an assortment of knives and closed up the hotdog stand. He and Jamie headed for home, but not before they had made a stop at the farmer's place.

The farmer took them out to the tree where he had strung up the carcasses for Ralph to dress out. After an hour of slicing and packing they put the boxes of meat in the farmer's cooler, and, having no place to wash up, they put the bloody knives on the dashboard, wrapped in a piece of newspaper. Ralph ahd Jamie were covered with splattered blood as they hopped into the small car and started up the hill to Brookins. They hadn't gone five miles when they saw a roadblock ahead and a half dozen policemen waving them over to the side. One cop was smiling as he approached the car, but when he shone his flashlight on the bloody faces of Ralph and Jamie, his eyes bugged out and his smile turned to a grimace. He pulled out a billyclub and his gun, then yelled, "Hey, Herb, we got the no-good murderin'. . . . Here's our bloody killers! Okay, you two, *out*! Real slow like. Hands above your head! *Move*!"

The policeman reached in, grabbed the red-stained newspaper, and carefully unwrapped the bloody knives. He revealed them to the other cops, who were encircling and handcuffing Ralph and Jamie.

"We're in luck. Look here! Here's the murder weapon."

"But Officer, you don't understand. . . ."

"Save it for your lawyer, Punk!"

Ralph never learned whether or not the rapist-killer he had been mistaken for had been caught. And if it hadn't been for the farmer's testimony, Ralph may never have been heard from again, except perhaps as the Animal Man of Alcatraz. . . .

8

Fifty feet of cable stretched ten feet off the ground between two trees. Zamba was clipped by his leash to the ring on the cable, and he was bouncing with increasing power back and forth, playing and exercising on the long run. Zamba was two years old now and weighed 375 pounds. He still had two years of growing to go but he already outweighed any of the other cats in the compound. He had a kind of arrogance about him, as though he knew he was a lion, but wasn't quite sure what to do about it. Often he assumed a superior stance and roared out through the valley, then flopped down and took great delight in the echo of his own voice. It was of course understood by everyone and everything on the ranch that Zamba held the position of supremacy here. Sometimes he had to remind Ralph just how important he really was and he'd devise ways in which to get his attention, such as the times he'd hide behind the tree. It didn't matter that the tree was only a foot thick and that Zamba's sides protruded way beyond the tree's, he knew that as long as he couldn't see Ralph, Ralph couldn't see him. He could hear Ralph saying, "I wonder where Zamba could be? What do you suppose has become of him?" And of course that was Zamba's cue to spring out from behind the trunk. "*Surprise*! Fooled ya again."

The surprise worked for Zam everytime, for Ralph always came over and threw his arms around him and scratched him under his chin (which he loved more than just about anything, except Ralph, of course). The way Zamba saw it, he and Ralph were two close male friends whose lives were all tied up in their masculine adventures with one another—of lion and man. Zamba also thought of himself as the young man's fierce protector, a chivalrous musketeer of the pet set. ("Listen, Kid, if anyone bugs you, anyone at all, you have but to speak his name, and I'll tear him limb from limb.") There was much pride in Zamba, as much as was in the young man.

This afternoon, while Zamba was actively displaying his physical muscle-building fitness between trees, Ralph was chopping raw meat for the carnivores' supper not thirty feet away from the barn. As the chain jerked around, it made a pleasant noise all its own, and Ralph knew as long as he heard that sound the lion was up and having a good time, usually throwing that enormous bone around or tossing his tire in the air or rolling his bowling ball back and forth. Zamba felt that he had a nearly perfect cublife. He owed his happy youth to this young man. He could have had a life of pacing behind bars—or he could have died by the river as his parents had—an orphaned cub with no way to survive on his own.

Zamba and Ralph.

Ralph had never left the lion alone on the long cable run. If he had other things to do away from the area he either took Zam with him or put him safely back in his pen. Now he stopped cutting the beef and looked around in a questioning, uncertain manner, eyes wide in a primitive instinct. Everything was too quiet. What had become of the song of the chain? "Oh, Zamba's probably just taking a nap," he thought to himself and began vigorously pounding away again. But no sooner had he started than he stopped and walked out the old barn doors to check his friend the lion.

"ZAMBA! . . . God! ZAMBA!"

The lion hung by the neck (the chain twisted tightly around it), dangling from a thick, low limb on the closest tree. His eyes were bulging; his tongue was limp and flopping at the side of his mouth. His two back feet were dragging the ground, the thick tawny body slowly swaying from side to side like a branch laden with heavy, ripe fruit.

At that very moment, with his mind so jarred by unexpected emotion, Ralph gathered in body tension, gained strength he was never aware he had. He ran for Zamba and in a muscular spasm lifted the great lion up into the air, managing to unclip the chain from his neck, then lay him gently on the ground. He began shouting for help. He put his ear to the lion's chest; there was no heart beat. Zamba's eyes were glazed and still. The cat was dead–he was dead. *"Not Zamba! Please! Please, God, not my Zamba!"* Ralph began beating on the enormous chest. He opened the great mouth wide and blew air into the throat. He pulled the massive forearms back and forth, then checked again for life. No breathing, no heart beat. He turned him over and pumped up and down on the lion's back. And just as Ross and Jamie and Bob arrived, Zamba let out a hissing moan. Someone got a blanket from the barn. They rolled the 375 pounds onto it and carried him over to Bessie, then drove him up to the house and put him on a mattress in front of the fire.

By the time the vet arrived, Zamba was breathing normally and just coming to. He threw up for a very long time, gagging and choking and having difficulty catching some air. When he was able to get to his wobbly feet, he was too weak to stand for any length of time and kept toppling over, bewildered, on his side.

Hour after hour he stumbled and fell. When he was finally able to stand, he walked, befuddled, in the direction of the calling boy-man. But as he started to move, he hit the furniture and stumbled across the room, only to collide with the wall. He could hear voices all around him, but try as he might he could not see. The boy cried out, *"He's blind–Zamba is blind!"*

Ralph could have endured that shadow for himself, but for Zamba to be rendered sightless was incomprehensible. Zamba was the essence of Ralph's character; he was all Ralph would have been if he were a lion, and he was as much a lion while there was Zamba as ever a man could be. Day in and day out he spent testing Zamba for signs of sight, refusing to leave him for blind.

The doctors explained, "You see, Son, the optic nerve, which is tubular, was pinched off during the hanging. Sometimes it pops back. Sometimes partial sight returns, and sometimes it doesn't."

Often Ralph sat up all night, pacing the room like a caged animal. "Not Zamba! Not Zamba! Not Zamba!" He'd groan in a raging temper, as though willing with his very anger the lion's sight to return. During the time of this blackout he declared war on the lion's condition.

"Optimism is a desirable state to live in, but sometimes it can deceive you," his friends warned. "Face it, Ralph, Zamba is blind."

That was like telling Ralph he was blind and would never see. One doesn't say things like that to Ralph, as I have come to learn. He does not know defeat. When things are at their very worst he will bounce back and change a devastating moment–one you may think in your faithlessness has no alternative but to remain a depressing negative. He will take that denial and turn it completely around. What you were convinced would have a bad ending he has the capacity to show you is only the beginning of something new and wonderful.

Every day of the next six weeks was begun with "Here, Zamba, here, Zamba, *come.*" And as long as the lion heard the boy's voice he could find him from his black and dismal forest–a journey of the intent hunt, searching for the light that once was, and had gone.

This sightlessness stirred a new terror in the boy-man. He had never conceived that any of his animals might die; much less had he given any thought that Zamba might go blind. He felt for those weeks as though his heart had been torn from him. He spent nights in hideous dreams–of having to put to sleep this fabulous lion. But in that seventh week of no dawn, he woke and called his Zamba to him. "Here, Zamba, here, Zamba, come on." And as he coaxed and cried, the lion saw a light, dim, in shadow, but a light. With each step he took, the sight increased, and as the boy called, the lion found him. To the left he found him, to the right, far away, and from very near.

The young man shouted, elated. He threw his arms around the great lion's neck, buried his face in his mane, and sobbed, "Thank you, God."

9

On foggy mornings Brookins was truly a magic place. The valley was covered in white, puffy, cloudlike masses, and the house stood out above the silver, elusive vapors like an enchanted island floating in a glistening sky. During such times of spellbound beauty, Ralph and Sahara would take their morning ride; then after chores were done in the early evening twilight, just when the shadows began to darken the ranch, he'd saddle her up again. His friends would do the same with the other six horses he now owned, and together they'd all ride off into the approaching night–many times when it rained, more often in the moonlight. It was during those pleasant rides, afforded by horsekind, that he was able to relax and fully enjoy all that he had, traveling the trails that always led to home.

The horses played an important part on the ranch, for they were the vehicles that carried everyone around. They lived in a picturesque barn that smelled of all the wonderful smells barns are so often known for–of hay and manure and grain—but above all the wonderful aroma of healthy horses. The barn was usually the first place the kids who lived and worked on the ranch came to. Some of them preferred the barn to any spot on the ranch. A couple of them even slept down there. Horses have a way of creeping into your heart before you even know how they got in. They get into your blood and become the better part of your dreams. I know, because I had a black friend once whose name was King—and his reign was a time of excellence—so perfect in all respects that I was disappointed to find other living things could never equal his kind.

On one of the days when ice and snow blanketed the land, when telephone lines were down and the roads were closed to town, on that fifth day without communication from the outside world, while they were all snowed in, a face came screaming toward the house—a face filled with anguish and agonizing despair—a face from a Goya canvas, wet with tears of horror and disbelief.

I ran to the window and rubbed the frost from the panes. Through a blurred glass I could see a young girl slipping and falling in the snow as she ran up the hill, screaming and hysterically calling my name.

"Ralph! . . . Ralph! Hurry! You've got to hurry! They're all dying!" And she fell and buried her sobbing face in her hands.

I ran faster than I knew I could, out the door and down the quarter-of-a-mile of mushy, melting, slippery road to the barn where Gracia's tracks had come from. Gracia was a sensitive girl whose frailty could scarcely survive any thing's pain, much less her own. She was tiny and delicate in the same way a folklore fairy queen might have been; you felt that if a wind were to blow her way she would lithely take to wing.

I threw the doors to the barn open wide and the steaming smell of straw and manure sailed past me. The snow had stopped falling, the sun was slowly rising, and morning was dawning gloriously bright and crystal clear. With every breath I took smoky mist was released to join that brisk, cool, fresh air. Just yesterday we were having snowball fights, turning the animals loose in the two feet of snow to run and play. Zamba was going berserk in the stuff, rolling and sniffing and jumping around like a cat with catnip—like a deranged clown.

Streams of light now filtered through the roof above and lit that hideous scene. Lying all around me were my horses. I couldn't believe it: Molly, Midnight, Son, Reble, Stud, Silver, and Sahara, horses that had been with me for years, had helped me to build this ranch. Dear God, the horses I had come to love. Some were dead, some dying, some swaying with weakness.

All the other kids who either worked on the ranch or just hung around had heard the screaming and now entered the old barn. Laurel, Jamie, Ross, and Don all registered the same shock I had just experienced. Each of us ran to the side of the horse we had come to know as such a close friend. Little Gracia was on the ground holding Midnight's head in her arms, soothing and caressing the dying animal, whose every breath was a gasping labor. We tried to get him to stand. "Please get up, Midnight, oh please get up for me, Midnight. Do it for me."

Pulling, tugging, bracing against our backs, he would stand. . . . "That's a boy, Midnight, I knew you could do it." . . . Only to topple over again.

We gathered blankets and made straw bedding, set up lanterns, and built a small fire, and everyone held the head of a dying child. God? God? God? What had happened? What was wrong? Had they been poisoned? Was it a disease? It was so heartbreaking and sad. As I went to Sahara's side I saw mold on the hay.

I had seen horses eat bits of moldy hay before, but they had never died from it. I pulled some of the fungus loose and stuffed it into a burlap bag, put some stool in a plastic container and a sample of saliva in a small bottle, then with a needle and syringe took some blood from my Sahara's neck and wrapped it in a towel for protection. All these things I gave to Ross and sent him on his way in old Bessie, whose spinning chains bit into the slush as they attempted to make it down the seventeen-mile "closed" road into town—to the veterinarian for help. We watched as he slipped and slid out of sight, disappearing around the mountainside, fearing that we might never see him again.

Molly and Silver were dead. We dragged their warm bodies out into the freezing snow. All that day we took turns sitting with the horses—working in shifts to clean the cages and care for the other animals. The meat for the carnivores was almost gone. We had been stranded up here for nearly a week. Today was the first break in the storm.

Next to die late that afternoon when everyone else was napping were Reble and Midnight. They died just as the fire went out. Gracia had fallen asleep with depression, her arms hugging Midnight's neck, her hands hidden in his mane.

Sahara was the last to go. It tore my guts out. I knelt down and brushed the forelock from her eyes, and for a long while I was mesmerized by the grisly scene. Then I cursed their dying and hit the wall of the barn.

A horn honked incessantly outside as Ross came skidding to a stop. I could see as I dismally turned that Bessie was piled high with food and equipment, and in Ross's hands as he came running was a bag of what looked like medication.

"I made it." He looked around at the sullen room and, swallowing hard, said, "It was botulism, Ralph . . . botulism. The vet said all we could do was put them to sleep. He gave me the stuff to do it with. Here." And he handed me the clear bag. "He said if we don't use the intestines and organs we can feed the meat to the carnivores." "Ross! What a thing to say!" "I'm sorry. But that's what he said."

Somehow Son and Stud lived through that terrible ordeal. Ralph had refused to give them a shot. Perhaps they had only a touch of that hay. Who knows? But they survived against incredible odds.

They pulled the three dead horses out into the snow and placed them alongside the other two, everyone cold and shaking, choking with tears as they strained and dragged the great masses, leaving behind their last impressions—a trail of dead weight. They would bury their friends as soon as the renewing storm let up.

Another day of icy cold and winter blizzard passed and another after that; then a third howled and whistled by. Nearly every animal on the ranch was brought in out of the violent weather's attack. Lions and tigers were chained to the living-room walls, chimpanzees and monkeys ran loose in the bedrooms, and every creature on the ranch who wouldn't fit in the old barn now slept in the warm house where a massive root burned in the big fireplace. The rest of the tree was still attached, and as it lay out across the wooden floor, it was thrust farther into the fire with every hour that passed.

Ross had brought back as much meat as the truck would hold, barely one day's supply. On the second and third days they fasted the cats, but on the fourth they had to face a grim reality. They could be stranded for weeks to come.

"We'll have to feed the dead horses to the others," Ralph said dismally.

"You mean butcher Sahara and Midnight and the others?" Gracia cried. "Never. Hunh-uh. No one's touching Midnight. No one!" she sobbed.

> I watched as their faces fell, and my own eyes lowered at the thought of plunging a knife into my Sahara. I wanted to vomit. Visions of the Donner party flashed before my mind, of their surviving by eating their friends, their family.
>
> At that macabre round-table discussion, as we looked around the room filled with pacing, restless anxiety, at animals who had not eaten for three days, we all knew there was no other way. What is death but the recycling of life?
>
> For a long time we stood by the side of the undug graves and tried not to measure our loss by those frozen, empty bodies whose souls had passed on. And as we hacked away at the flesh of our friends, tearing open those walls, and when I saw little Gracia inside the open cavern of Midnight, marred by sorrow and bloodstained tears, butchering the horse she loved and allowing no one near, I felt as if my heart would break. I thought then that the end of all I had come to know was very near.

10

In the spring of 1955 Ralph made his way from the mountain fantasy of an animal lover's paradise. His new search had found a lovely beige ranch resting just on the edge of the high desert, in Agua Dulce Canyon.

Leaving the Brookins estate, he felt that a part of him was staying behind. It seemed a strange, sad loss that he would never again drive the green-wooded Malibu roads, never again gaze down on that beautiful lake. He resolved always to remember the three years he had spent there as a milieu of enlightenment, of times spent in deep and satisfied thinking. It was there that he had celebrated his coming of age, had found humility, and had considered his fate with a new view. Now a small chain of events linked him with a past. He could momentarily consider the importance of looking back, and by so doing he saw what the future held. His experiences there had banished all boyhood vanity. This new insight made a man of Ralph and to a certain extent leveled his world of ideology, not influencing the goals of his own righteous mind but those his ideas planted in the consciousness of others.

And it seemed sometimes he rode on a magic swing, which, when thrust forward, gave him a brief look at the perfect illusion, then, in a free-swaying motion, brandished him in a manner of great horror and pain. And as the pendulum swung slowly to a stop, each extreme feeling diminished until he was caught quite still somewhere between heaven and hell. This appeared a fair analysis of life to him—that as long as he could stay on the forward swing, in all ways ready for that flight, his faith would carry him on, and all he envisioned as what should be, would be, if he didn't look back.

I have chosen one story from that Agua Dulce period to establish a truly great character in our history. She is equal to the writings of Richard Adams, but to date will have to settle for the likes of me and Ralph.

Ralph's first big, profitable job came with a call from Desilu Studios requesting in particular an elephant as well as a variety of other animals to work the full run of a new TV series, *Frontier Circus.*

"Of course, I have an elephant," he replied, and made the arrangements for the coming interview. He hung up the phone, took a great sniff of air, and erupted into vocal exaltation. Then a brief moment of panic seized him.

Not only didn't he have an elephant, but even if he found one on that short notice, how could he possibly have it trained in three weeks? He made dozens of expensive calls all over the United States. A few elephants were available, but their price was outrageous: four, five, six thousand dollars. (Would you believe that today it is thirty thousand dollars!) He had, as usual, little excess capital. Economics forced him not to consider more than $2,000. A bit of good luck came when one of his contacts informed him that an elephant who had lived for some years in a zoo in Tennessee was for sale. How many zoos could there be in Tennessee? Ralph called them all and found his elephant. The price was right: $1,500. He bought her sight unseen and was beside himself with the excitement of his very first encounter with an animal he'd always considered kin to man.

> We ran toward the approaching semi, which crept toward
> the compound, swaying from side to side like a rhythmic

cradle. Bert pulled her to a stop, and a billow of soft dust settled down, draping us all in a thin grey coat. The trailer continued to rock gracefully, creaking to the music of its own steady pace. We threw open the back doors with eagerness. The combination of moist, hot air and the strong aroma sent us back a few steps, rubbing our eyes and gulping fresh air.

It was twilight, and the last rays of day shone through the waves of rising heat, lighting the elephant, who stood pedestaled, towering above us all as Turner would surely have lit a golden, romantic subject beneath a burning sky–a subject to be worshipped in the midst of flaming color, shimmering and flickering, radiated by glow.

She was nearly as big as the trailer was tall. But my dramatic vision began to fade as Bert brought her slowly forward out of the truck, into the last bit of light. She wobbled on tall, thin columns that barely supported the sagging gray body. Her trunk was not like that of other elephants. It hung long and straight and dragged the ground like a lifeless limb. Her ears dropped loose and forward. One eye was white and clouded. I moved closer to inspect the enormous hulk of ancient decay, and Bert, who until now had remained silent beside her, holding the pink, freckled wafer ear that occasionally attempted to fan, shouted at me, "Don't go near that side of her Ralph; she's blind and shy on that side."

"Blind! Blind, Bert? I don't believe it." I stepped to the side of her good eye and touched a tuskless face. A rumbling issued, deep and low, calling like the sound of long, lost thunder in search of a storm. She looked down at me from her second story with a gentle, frightened eye; the cavity that was her mouth opened, and a great fleshy tongue, swollen and sluggish, showed itself. I petted the wet thing, and she leaned into me.

Blind, a paralyzed trunk, half a ton underweight–not to mention that she was sixty if she was a day. Somehow this wasn't exactly what I'd had in mind.

"Well, you were wondering why the price was so cheap; now you know. Ralph, meet Modoc." Bert cordially introduced us.

Later that evening after we'd bedded Modoc down, we sat in skepticism beside the fire in the cabin, filling out a chart regarding the elephant's history while it was still fresh in Bert's mind. But there wasn't much to tell, and so mystery surrounded her. For nearly ten years she had been chained to a post in the zoo, where she shuffled about in a life away from other elephants, a life made of little rituals to get through the boredom of her day–eating, drinking, going through the motions of being an elephant. Her keeper, who was petrified of her, treated her as one does the things he fears–roughly–misunderstanding her, nervous. He told Bert that a circus had dropped her off some years back when her handler had put a bull hook through her eye. He had been told she was a potential killer because of her eye and her age and to stay clear, which was exactly what he did. So for all those years she had remained barely alive, standing in one little spot, swaying to and fro to keep the circulation going. Now her bones were so stiff she could barely walk. I wondered what circus had abandoned her with little sympathetic consideration–a senior citizen of the animal world, cast aside with age and placed alone in an unfamiliar spot to erode.

To awaken Mo from a ten-year stagnation was not an easy undertaking, but we fumbled along. I had practically no experience with elephants; so in a way I was fortunate that old Mo was in need of an overhaul. Her age had somewhat mellowed her. We gave her a choice of food, and as Jerry Lewis would have said, "When Modoc ate, L.A. rehearsed for a famine." Incredibly, we had her ready for the series I had contracted her for. In less than one year she stood her full 7 feet 10 inches and weighed 9,180 pounds on the scales at the truck stop–nearly five tons.

Her steps were small at first, but with each day the old joints became more supple, and she moved a little faster. Her strolls became longer. It was like taking a building for a walk.

As her new-found energy increased, the blind eye took shelter in me. When we exercised, I worked her only on her dark side. By doing so, I became her seeing-eye person. She was secure while I touched and talked to her. But one could never–ever–walk up unannounced on that blind eye, for to startle her would result in a frightened, intense swing, an uncontrolled but accurate whip from her numb but still useful trunk. And the forgetful person would sail head over heels, astonished at having forgotten.

Old Mo lived beyond the ordinary level. She emphasized the need for a sensitive world, one measured by tolerance. She was a magnificent elderly lady.

Three Hollywood stars–Modoc, Chill Wills, and Thelma Ritter.

The story of Modoc came out when the retired circus performers held their annual picnic at my place. They were a marvelous combination of special oddities–members of the old Ringling band, clowns, acrobats, roustabouts, aerial artists, trainers of all kinds, fat women, small people, tall men, tight rope walkers–a warm-hearted profusion from out of the past, who sang songs and reminisced for one day each summer when their circus came to my town. . . . Yesterday. More's the pity for no more yesterdays. . . . An intensely human group, who had come through–made it this far–hugged and kissed and congratulated one another at having survived, then sadly counted those usually among them who were missing and noiselessly vanished into their reunion.

I went to the barn to prepare Mo for the little performance she and I were to surprise them with. In the seven months that

Mo, the world, and I had been one, she had recovered. She was alive again, no longer hastening to her end. Fat, agile, magical, she was a fountain of enthusiasm and energy. I had to move to keep up with her now. After brushing her off with a broom and before leaving the barn with her, I ran back to the assembly of soft-sound effects and announced that I had a few minutes of entertainment for them. They clapped with velvet hands and smiled toothless grins and graciously took their seats about the picnic tables as I checked the "curb ring" and placed its opening panel in order.

"Oh, elephants," one happily declared to the other.

"Only one," I answered back.

Everything in position, I went back for Mo. She made her entrance as though she had sprouted wings and bowed low and long on one knee as I introduced her. "Ladies and Gentlemen, my friend Modoc." The mention of her name created a wave of whispers flowing in and out. I heard someone say, "But she's dead!" And another said, "It can't be." And another, "But look at her trunk; look at her eye!" And Mo and I gave nothing less than a great performance in the old circus style. When we finished, a hearty cheer went up, and we received, of all things, a standing ovation! (We were good, but we weren't that good, I thought.)

"Does she dance?" someone yelled.

"I don't know, let's see." And I said, "Dance, Mo." With that, the mountainous sides began to shake, and the feet began to prance, and the distant hills were pebbles under her beat.

"That's her!" the same voice yelled out. "That's Dancing Modoc. That's the Golden Elephant!" And she and I were suddenly surrounded by wet, shining eyes. Friendly arms embraced her trunk and patted her sides, and she rumbled as her name was said over and over again. How priceless a piece of the past.

The Golden Elephant of the Ringling Brothers 1933 golden anniversary celebration, who had held center ring for so many years as the famous solo-dancing elephant, was none other than my old Mo. And they recalled. . . .

My famous Modoc had been with the circus since she was seven years old. Her career had begun in 1904 and had run a span of half a century. She had lived through the horrible poisoning of the Ringling herd, when a dozen of her stricken friends had died and left Modoc terribly ill. She had survived train wrecks, fires, storms, angry mobs, one-night stands, and a life far from her native land. And they remembered more. Her trunk had been paralyzed from pushing heavy circus wagons around the country. A drunken handler had beaten her and put her eye out with a bull hook, and the last they had heard she had died.

Ten years later we had found a lonely old one-eyed elephant shackled to a post in a dirty, rundown zoo, and all the years that had gone before her had been forgotten.

But they remembered Mo. And we'll never forget.

It was at the Agua Dulce Ranch, by a twist of fate, that I had met Ralph, who by then had, as I saw it, the understanding of a sage and was heroic in every outward implication. He was then–and remains–the only man in my life. It has seemed a very natural thing that I should love just one, for the one I chose set such a high standard–far above and beyond the reach of another.

3
Wild Eyes - Gentle Hearts

There was never a king like Solomon
Not since the world began
But Solomon talked to a butterfly
As a man would talk to a man.

–Rudyard Kipling

1

Ralph had been gone six months, and my misery seemed more than I could bear.

He returned in April, and in May I took him to Riverside to meet my parents. I hadn't really prepared them for this extraordinary man who was about to walk into their ordinary, middle-class lives.

Stevie opened the front door when she heard the car pull into the driveway, and with her usual theatrical flair she cried, "Good evening, Bwana, and you too, Toni."

"This is my younger sister, Stevie. Isn't she sweet?" I said through clenched teeth.

She pulled me back and whispered into my ear, "Let's not be bitter; so far you're holding your age well, Dear," then practically shouted so everyone could hear, "Now I know what's meant by the call of the wild. Does he have a brother?" Ralph was beginning to show signs of discomfort.

Stevie bowed low and blithely announced, "Destiny is waiting," and led us into the living room where bourgeois intellects waited to pronounce judgment. I was experiencing hot flashes of uneasiness because of the baited trap I had willfully led poor Ralph into. If tonight proved to be a disaster, it would be my fault.

Both my parents stood when he came into the room. I introduced everyone, and Ralph presented my mother with a box of See's candy and my father with a bottle of Lancer's wine. We sat through moments of awkward silence until my mother said cheerfully, "Well, Ralph, we've heard so much about you and your unusual life." Why did she say that? I had hardly mentioned him to my family. I was a big girl now; I didn't have to report in. "In fact," she continued, "I was just reading an article about you in the Sunday parade section of the *Press-Enterprise.* It says here you were paid $75,000 for your participation in a film. Is that true?"

"Mother!"

"Toni, dear, why don't you and Stevie go clean up your room while your father and I have a little chat with Ralph?"

"Mother! I don't live here anymore, remember? And besides, I'm not leaving this room for one minute."

"Very wise decision," Stevie whispered.

"How are the bear cubs doing?" asked Scotty.

"Just fine, Sir; they're quite large now. You'll have to come out to the ranch and pay them a visit."

Before he could say another word Stevie interrupted with, "What's Bill Holden really like?" and so began a half hour of film star trivia: "Do you know, and is it true, and does she really?" Ralph held the floor with movie gossip, and I knew he hated every minute of it.

My mother brought in some little sandwiches and fruit juice and I began to relax as I could see he had the situation under control.

"So you're a lion tamer," Scotty said abruptly.

"Well—yes, I work with lions, but—"

"What he does is a highly technical and extremely complex behavioral science," I interrupted, hoping against hope that Scotty wouldn't ask any more questions.

"I was addressing Ralph, dear. How old are you Ralph?" he continued. Ralph, who sat next to me, squeezed my hand and smiled as if to say, "It's all right, don't worry." "I was thirty last week."

Before the conversation could go another step, Stevie had brought out my scrapbook, placed it on his lap, and turned immediately to the pages of my old boyfriends.

"And here we have the past she left behind. This is Bob," she began.

I grabbed the book and Stevie by the arm and we made our exit in smoldering silence to the kitchen, where I planned to strangle her.

"Do you have to keep embarrassing me?" I demanded.

"Well, rules are rules," she responded.

"Are you lonesome for agony since I've left? Because if you are, I can certainly give you some, such as–Mother, did you know this naughty little child has not been staying with me on the weekends as she has led you to believe? Instead she has been organizing a rally to march against sexist oppressionism on campus, and she is about to expose the Chancellor as the head of a fanatical religious cult–and–"

"You wouldn't!"

"Oh no?"

"That's blackmail."

"That's right."

"It figures. All right! All right! I suppose one must expect inconvenience to some extent even if it is in one's own house. I'll be good."

"How can I ever repay you for your kindness?" I twinkled.

"By leaving," she insisted. "All this torture has made me thirsty."

I filled up a glass with foggy water from the tap. "Here, have a drink of bilge."

She swallowed a sip and ran screaming from the kitchen–"Mother, she poisoned me!"

"I stand accused," I said defensively, "but I did it for the good of the family."

"Do you suppose he would like to see my imitation of Kate Smith?" Stevie asked.

Ralph had an odd expression on his face that said, "Where am I?" and "What is this place?"

"So you're nine years older than Toni," my father declared. "That's quite an age difference, isn't it?"

"You're nine years older than I am, Scotty," my mother interceded.

"Well, that's different. You were twenty-six years when you married me. Toni's only twenty-one; she's still a baby."

"Marriage? What are you talking about marriage for?" I gulped. "Oh Daddy, how medieval." Stevie sprang in, and they began to argue about the importance of maturity in a man.

Poor Ralph had hardly been able to say a thing the whole evening. When we left, I had the impression that he was bewildered.

"Well, what did you think?" I asked.

He grinned, took a breath, and replied, "It's a nice place to visit, but I certainly would never want to live there."

I called my mother the next morning. "Well–what did you think?"

"I think he was charming, perfectly charming." She hesitated. "For someone so unusual. But, Dear, your father and I have been discussing the way he lives. You know, Toni, that's a very dangerous business. Don't you think perhaps you're more attracted to the romance of what he *does* rather than to what the man *is*?"

That was a very interesting question, one I would have to consider in depth. I came to the conclusion that there was nothing so bad about enjoying the profession of the man you love. In fact, it seemed to be a definite advantage not only to love him but his work as well. I mean, I don't think I ever would have been attracted to someone who designed funeral attire. What a person *is* is reflected in what he *does.*

While Ralph had been on location, the county had begun to chop through the beautiful old Agua Dulce ranch, dividing the property in half, scarring the landscape to make room for the new Palmdale-Lancaster Freeway. So Ralph made plans to purchase another piece of property, not far from the sleepy town of Acton. With the money from *The Lion,* he bought what was soon to be known as the famous Africa USA.

In August of '62 we eloped and were married without due formality. Neither of us wanted a church wedding, and although it was against my nature not to discuss such a major decision beforehand with my mother, this was a time of splendid and unequalled beauty in my intimate life, and it did not seem appropriate that Ralph's and my plans should be discussed with anyone.

We kept our marriage to ourselves for several selfish weeks. When we announced it to the world, it was because the burden of such happiness was now too heavy for us to carry alone.

My sister's comment when we told her we were married was, "May the pterodactyl of happiness fly over your wedding cake." When my mother visited the ranch for the first time, she said, "I know all these people, Dear. Why, they're the ones who never danced at the Prom."

And my father was known to have said over and over, "All that money on her education, all that money on her education, allthatmoneyonher-education."

Our little house, although small, was quite comfortable. It sat snug against the mountain, and in August the tumbleweeds piled high against our back walls. The view from my kitchen window was of a steep incline covered with sage and teeming with crawling things that occasionally greeted me on the porch of my front door. We wore boots here most of the time. There was no grass, no flowers, no planting or growing of things. We were too busy for that. All our efforts were concentrated across the street at the compound where an overabundance of life was reproducing at a rapid rate.

The evenings were the nicest part of the day, when, after feeding time, the animals began to chant and celebrate with song, a melody in which so many sounds were sung–the bellow, the trumpet, the roar, the howl, the cry, the wail. And then once again the reply as the answers came in response. And early in the morning we were awakened to the same pleasant noise. I cannot imagine our lives without these sounds.

Saturday night barbecues and barefoot dancing under a string of paper lanterns, picnics and songs around blazing campfires–these things added zest to our lives. The move was wonderful for me. It seemed as though I had come in at the beginning, and Ralph and I were starting out together. Everyone at the compound seemed to have adjusted their lives to fit the animals. The majority of employees lived on the ranch and worked long, hard, devoted hours caring for the animals they loved, who in return loved human companionship.

Africa USA reminded me of a mini Peyton Place centering on a male-dominated community, most of whom had been following Ralph around for years.

Unfortunately, many of the Helfer squatters disliked humans as much as they loved animals. At times I felt like Napoleon in exile. Friendship was difficult to cultivate in their little company; and of course I never told them of the fantastic opportunity they were missing by being rude to a really swell person–me!

Ralph was not only the Mayor of the town, he was the entire municipal governing force, not to mention the family "shrink," plunging into the shallows at least twice a day. A man who has always sympathized with others in pain, Ralph attempted to get the mental processes of his employees functioning at a close norm. He felt a parental responsibility for their continuing happiness. But more often than not he simply became a prisoner to their confessions!

I will take this arbitrary space to introduce you to a special member of our cast who played an important role in our crazy life.

At night when the rest of the world was asleep, Ralph and I would stroll about the park as other nocturnal animals do, acquainting ourselves with the dark side of nature, which usually gave me a slight tinge of the fantastic and other discomforts too numerous to mention.

The freshness of evening air and the strong smell of life coupled with somber illumination and the reflection of a twinkling sky upon our land was heightened by sounds from the unknown; and though we practiced the art of peace, our minds were often ready to do battle against the unidentified shadows that sometimes hid behind the flowering creepers, following us down mysterious paths and holding over us an invisible, imaginative power.

Working for us now was an ex-convict, who, having served his time for stabbing another man, was released to renew his crimes. And Ralph, believing the world innocent and society guilty, gave the man a helping hand by signing a document making Ralph responsible for him—assuming custody of him. The man became passionately and forever in debt to Ralph for believing in him and so set about to prove to us his steadfast loyalty. He duly appointed himself as our guardian. We could not imagine what it was he thought to guard us from, but like the African askari, he showed up in the most fortuitous places.

Ralph and I walked through the park about 9:30 one lovely evening, occasionally stopping to check the locks on the cages and bid the inhabitants good night. We became suddenly aware of a steady movement of foliage all around us, and as we increased our speed, the brush moved in a current with us. Our hearts pounding in feverish breathlessness, we started to run and were stopped dead in our tracks by a tiger that appeared to be considerably larger than those we had known. As he stepped out onto the dark pathway before us, neither Ralph nor I could tell whether he was friend or foe. We puffed and received puffs in return. Now frozen in our tracks, we began to look for cover, where we could plan what to do. Then I heard a voice—mine: "Nice Kitty, that's a good kitty."

"Ralph, who is that?"

"I don't know." We backed up to a tree, and just as the tiger was about to spring and I about to scream (you have never heard such a noise as comes from my throat when I am under attack), who should appear from the cold darkness but our shield from all danger, the ex-con. He stood like David before Goliath, between us and the cat, and as the animal leaped to the spot, the ex-con wrestled him to the ground, where a leash (which he just happened to be carrying) was placed about the tiger's neck. He called out, "Don't worry, Kids, I've got this situation under control." And the laughing, puffing feline was escorted to the empty cage just beyond at the end of the Wild String. The tiger was Sarang, star of many films and a worthy comrade. But in the dark of a pale, clouded moon our dissolving anxieties had denied our recognizing him.

We exclaimed admiration for the appearance of our guardian angel, and he blushed and shuffled his feet about the ground. I knew Ralph thought him to be magnificent in his shyness, for of course here was visual proof of his atonement. He was a proud sight in his self-sacrificing way.

Naturally, we derived a parental pleasure from telling about the man's heroics. But when similar confrontations with loose beasts in the night happened again and again, and to our side each time came the paradoxical ex-con, we questioned the multiple acts and proved them too similar to be coincidences. It appeared that our ward drew great pleasure from saving our lives and so set about designing romantic swashbuckling scenes where the child in him fulfilled his desire to rescue his family. After any misguided interpretation others may have wronged him with, having now abolished all those abuses and releasing his unjust guilt, he could walk as a hero, proud among men. This act of eliminating the "bad guy" in him was his way of conquering his own distortion, but Ralph and I could not continue being saved from his defenses. In an attempt to rechannel his energies, we discussed with him our suspicion about in-house crime, and we sent him to investigate. Several weeks passed before he asked to speak with Ralph. It would seem that our detective had found the guilty party, and, believing the nameless person in question was not an evil doer, but, as he reported, a "misguided individual who swore it would never happen again," the ex-con, never revealing the man's true identity, had taken it upon himself to assume custody of the man's ailment. Once again he emerged as a hero as he attempted to right a wrong and atone for a delinquent deed.

We were never set upon again. Instead, young animals began to disappear, and we called in the police. But we never found a clue as to their whereabouts. Guess who solved this puzzling affair? None other than our forever savior, the ex-con. Lo and behold, the stolen animals suddenly reappeared, and we were requested not to ask any questions (honor among thieves, you know) but to be thankful our very own convict knew friends in the underworld circles who had traced the whereabouts of the kidnapped victims and captured the assailants, dealing with them on their own terms. For any friend of the ex-con's was a friend of ours.

He stood glowing, pleased as punch, holding two stolen mountain lion cubs in his arms. The victims of his pride once again, we thanked him for saving the day. But when Clarence the Cross-eyed Lion disappeared, that was too much. Upon the return of the ex-con, sporting at his side one bewildered lion movie star and while he was preoccupied with tales told from his vantage point, we unmasked his disguise, refusing from that moment on to feed his dependence. And so we exposed the vulnerable romantic. In his defenselessness, he shrank with shame. All pretense ceased after our electric-shock treatment, and he confessed with agonizing disclosure to the crimes, then violently pleaded for absolution. Now he became the confessor, and we listened for hours to his long list of disguised sins. Having been imprisoned by his now-relieved burden, he had shown great courage, it seemed to us, in admitting his inadequacies through this brutal exposure. We therefore verbally defended the ex-con's other nature and supported the mature human elements. He rose to elevated heights at the consciousness of having been championed. He had gained some new kind of high through the extension and expansion of his own confession. A mantle of self-embodied protection arose; we witnessed his self-vindication. And so he exonerated himself. What do you do with a man like that? In many ways it was wonderful working in such a madhouse.

From the trains, sometimes rag men were thrown, and they made their way from the railroad tracks to our home, dirty and tattered, old and worn. Bearded and red-eyed, they'd knock and ask quite civilly for a spare bit of food. We often entertained a nomadic wanderer, since Ralph took under his wing all kinds of lonely things. And some stayed more than a day, while others ate and went on their solitary way. It was not unusual to find a hobo sleeping in the tunnels under the tracks. And I think more often than not those who arrived in a stupor awoke to the surprise of jungle calls in an alien world where scavenging had become a long lost trait.

There were other kinds of vagabonds that dwelled here as well. They lived in caves and burroughs beneath the sage and underground. At night, with their their yellow eyes, they came out of the hills and roamed the ranch, taking food without permission, at a stealthy pace. With the first light of morning they were gone, having left evidence of themselves be hind: footprints of coyote and mountain lion, of bobcat and deer and bear. How sad that as I write now the mountains have been cleared of such things. But in my day (not so long ago at that) wild animals spent their time freely here.

Old Chester Higgins was our closest neighbor. Living with his feeble mind landlocked just at our property's edge on a rise not much bigger than his house, he ranted and raved and carried on quite madly for the eight years we were there. He was short and pot-bellied and his unusual head was in the shape of a geode. His eyes were small and receding, his nose rather hooked, and his mouth offensive. Sometimes at night when I was checking to see that everything was right at the ranch, I'd hear Old Chester howling with my wolves and then laughing quite insanely at the top of his voice as though followed by the hyena, and that wildly foolish madman would, on occasion, pop out from behind a bush he had trespassed beside and give me a near seizure. I thought more than once about shooting him to put me out of my misery.

As far as Higgins was concerned, the even older Reverend Dooley who lived just half a mile to the east in a pig pen was reduced to the state of slavery. For by all rights, since the reverend couldn't prove he was a "freeman," the black gospel preacher belonged to him. And so long as Old Dooley lived in accord with what Old Higgins delegated, he could by his claim squat there. Inferior, discriminated against, abused by prejudice, the old man was treated by his neighbor as an oppressed captive. But the Reverend, singing a hymn and spouting the ways of the Lord, was inspired by divine guidance to save the soul (at the bottomless pit) of Chester Higgins, whose condition of heathenism demanded his undivided reli gious attention. The wordy combats those two antiquated, weatherworn fossils engaged in so vigorously, one with withering scorn and one with saintly grace, were apparently enough to kindle, on occasion, the otherwise inactive fuel; for from where we sometimes stood, the loud contention was visible–Higgins's expressive finger flying not unlike a banner as it pointed with four-letter vengeance at the African missionary, who called on his Christian God to guide Higgins in the way of Jesus and "let him be shown the light."

The "nigger" and the second class citizen were in their way good for one another since their relationship gave each a cause to fight for and a reason for living in the otherwise empty world of the abandoned eccentrics. So they found a special kind of friendship that heightened each of their spirits and gave them something to look forward to–if only a heated argument.

2

My house was not the usual kind of home. When we were married I knew little of wild animals other than what I had read in books, and most of those were filled with limited knowledge. But as little as I knew about mammals, there was one thing I knew quite certainly about reptiles: I had a pathological loathing of them. When I moved in, I said to Ralph, "You must have a lot of tools to keep so many chests around." Ralph smiled but did not respond. "What kind of chests are those?" I asked naively.

"They are army coffins."

"Army coffins! Oh, how original. You keep your tools in army coffins." There were locks on all the coffins.

"You leave the front door open out here, yet you lock these boxes. Why?"

"Ahhhhhh."

"Those are tools inside, aren't they?" I enquired quizzically.

"Yes—you could call them tools—of my trade," he answered.

"Why are they plugged into the wall?"

"Ahhhhhh," he tuned out.

I allowed him a wide range of responses, yet he remained elusive. I had a sense of apprehension, but realizing that I was on probation and that my good behavior in these beginning weeks would probably have a positive effect on my marriage, I decided to let him tell me his little secrets in his own good time. Besides, as soon as he went out of the room I could peek into the holes lining the side of each of the four coffins and see what, in fact, he had in there. He stepped from the kitchen and with interest aroused, I tiptoed toward the boxes.

"Toni!" he shouted, having sneaked up behind me. At that very moment I was ready for a coffin of my own. Recovering my wits, I turned around, still embarrassed at having been caught trying to sneak a look. Then I thought better of the whole situation. This was my house, too.

"What ever happened to loyalty, integrity, honesty, and sharing with your friends? How can you say you love me and still keep deep, dark secrets? Am I not entitled to know what's going on in my own house, for heaven's sake?"

"Ahhhhhh," he whined.

"Will you stop saying that? Trust me, I won't tell, no matter how awful it is. I am not here to judge or punish you. I am here to listen and to help, if I can," I said with maternal compassion. "And besides, a wife can't testify against her husband."

"Honey," Ralph began shyly, "you're right, you are absolutely right. Let's just sit down in the living room and have a little chat; then we'll open the coffins, okay?"

"Ohhh kaaay! . . . But shouldn't we wait until after dark to lift the lid? Can I ask you one question?"

"Sure."

"Is there dirt on the floor of the coffin?"

"Dirt?"

"Yes, you know, earth from home—haven't you come to give me Eternal Life?"

"Good grief, Toni, what do you think is in there?" He was a born straight man.

"Well–nothing right now."

"Shall we continue?"

"Please."

But before he could continue, an unusual noise filled the air–not the sound of a happy ending–no, this was the sound of wood splintering.

"What was that?" I asked.

"We shall see," he answered calmly. "In the interest of safety I have decided to check it out."

"Oh, thank you."

Ralph jumped up and ran toward the kitchen. I followed closely behind.

"Too bad," Ralph said.

"Too bad about what?" I asked.

"Too bad I wasn't able to explain first."

What was he talking about? My whole family is into one-liners, but Ralph's never made any sense at all.

We now stood before the shaking box that seemed to be coming apart at the seams. The top was expanding.

"Show time!" Ralph called out.

The coffin slowly began to open, and before my bulging eyes there arose from its depths the Loch Ness Monster. I couldn't move. I was paralyzed with fear. Why, the "thing" (I use that term loosely) must have weighed *200* pounds. I would have preferred a confrontation with Dracula. Ralph patted the giant stalk (who was now scanning the room like a submarine's periscope) on his head.

"Well, as long as you're out of the box, 'Hisster,' shall we meet Toni?" Holding six of the twenty feet in his hand, he had the nerve to come toward me! I began to shake like a Stauffer's reducing machine. Then I screamed and ran outdoors.

After a great, pleading apology on his part (of which I loved every minute) I threw a tantrum and insisted Loch Ness and her string of friends be removed to the guest house.

Now, under normal circumstances I would never issue an ultimatum, but at the moment I didn't consider either these circumstances *or* Ralph normal, and under *no* circumstance was I about to live in my house with giant reptiles. Out they went, and, thankful and joyous, I returned.

One week later I heard myself saying, "I don't think so–I don't want to–Thanks, I'd rather not. Must I? Oh no, why me?" Although the student was clearly hesitant, the inspired teacher had nonetheless arrived, and

Ralph and friends in Africa with the "Loch Ness monster" and Capucine, Pamela Franklin, William Holden, and Trevor Howard.

together we began my adverse journey into the world of reptiles.

"I hate it, I hate it! . . . It's obscene!"

"Are you afraid, Toni? Because if you are, just say so, Honey, and we'll start by breaking down your preconceived ideas, unless you would prefer not to be involved with reptiles at all. I'll understand; I'm only sorry I didn't realize sooner you had this problem."

"Afraid? Me afraid? Let's not be ridiculous. Hand me that thing!"

To begin to unravel the riddle of serpents in his Eden, he gave me a shiny black indigo. Each day I was to handle him "carefully," never, ever, to drop him, since to fall was a snake's greatest fear.

"He needs security against his acrophobia."

"Oh, how adorable, a snake with a phobia." I thought that was terribly funny, because there wasn't a name for what I was feeling–only a shivering sound.

There I sat for days with a twinge of tabloid mentality, passing time staring at a neckless, chinless, armless, earless, legless torso and thinking, "What if this thing didn't crawl on the ground? What if it were to stand up and walk toward me, or worse yet, what if it could fly . . . ?"

I kept remembering the *big* snake story–back in the Helfer archives.

Ralph had become a stuntman in the 40s, filled with blue-eyed innocence, examining his latest wounds with humility and a sense of proportion. There were occasions when the cinema demanded more than he knew how to give.

When stunts were called for in Hollywood that no one else was willing to do, they called Ralph. But when he was asked to double as a rather voluptuous female, the suggestion seemed so outrageous that the prop man, even as he asked, couldn't keep from laughing. Somehow Ralph Helfer as a female impersonator was a vision that had never presented itself.

On the day he was to wrestle a 24-foot, 240-pound reticulated python, he was dressed in feminine frills, bulging with biceps, protruding with falsies, and cinched with a girdle. It was certain that when he walked out of wardrobe he would create mild hysteria. But he was not prepared to receive a box of long-stemmed roses and a love note requesting his presence at dinner that night. He spied one of his male assistants holding his sides and running from the sound stage. And as if that wasn't bad enough, he was assigned an effeminate male choreographer who was to teach him how to walk with "style and ease." The dancer's obvious attraction to Ralph set his teeth on edge.

Poor Ralph was beyond consolation. Casting haughty glances about the room at the smiling faces, he placed the rubber mask that was the star's likeness over his head, and the hairdresser attached a long, flowing wig.

Ralph had no big pythons of his own then; so he had borrowed one from our good friend and renowned herpetologist, Jim Danielson. This python was not among the nicer snakes who crawl the earth, and Ralph was not assured by the fact that she had to be dropped onto him from a tree as he walked under its branches, since it took five men to pick up the snake.

They rolled the cameras as he walked mincingly out onto the jungle trail. When the mammoth weight hit his shoulders, he fell to his hands and knees, where he remained for a few unbalanced moments, then gained his footing and stood, already exhausted from the exertion of supporting 240 pounds. He was drenched with sweat that poured, burning, into his eyes behind the mask. He took savage gasps in an attempt to take in an adequate amount of air and tucked his chin to keep the snake from coil ing around his neck.

As he did so, the movement of the snake twisted the mask to the side of his head, and he was instantly without breath. He couldn't scream, he couldn't see, he couldn't hear. He collapsed to his knees, suffocating, rendered powerless. The great coil then slipped an iron noose about his neck and constricted. In sheer panic and fighting for his life, he completely forgot the sign language we use in moments like these. Each hand signal meaning, such as "I'm all right; how much longer to go?" "Take him off, *help*!" Certain he was about to die, he could not summon even those survival instincts. The worst part was that no one knew what was happening to him because they could not see his facial expression under the mask. His heart beat as though it would burst through his chest. He suspected everyone could hear the loud rapping as he fell into unconsciousness and dropped the great serpent's head. The python again constricted, taking his last breath, waiting for his heart to silence.

When he released the head, his men, realizing he was in trouble, rushed to his aid. By then Ralph had been squeezed into a limp coma. As the men tried to approach him, the snake, holding him in a possessive muscle, held them all off. She assumed a striking position and lunged with over 200 hypodermic teeth (all pointing backward), threatening those who moved her way.

Then a very daring young man crossed his face with both arms and in complete defiance of the danger swaying before him threw his body at the great snake and was severely bitten in the abdomen. With a bold and courageous effort he grabbed the head, which held his stomach in agonizing pain, as the others dashed in and unwound the python from Ralph's lifeless body. They tore the mask from his face and administered artificial respiration.

At the hospital he was given oxygen and injected with heart stimulants. When he regained consciousness, he was so nauseated he actually hoped to die. That he had been brought in dressed in feminine clothing aroused a good deal of suspicion.

3

Besides the snakes, moving into a house with a full-grown African lion was something I had not been previously schooled for either. It would seem my expensive education, the one my father constantly reminded me of, was in fact sadly lacking. Zamba, Ralph's benign friend, lived in the breakfast nook to the left of the kitchen. He had an accommodating outside area attached to his room, with a lavish kitty door for entering and exiting. His biting smell was hypnotizing as well as overpowering. It also smarted the eyes, and I often had to cook in competition with it, usually with a hankie over my nose.

Life among the lions was subservient. My first duty on rising was to clean Zamba's sandbox. My early morning harvest of lion spoor did not start my day off well. Zamba gave me the definite impression that I was a guest in his house, that as long as I didn't step out of line, I would be temporarily welcome here. I had the suspicious feeling that I had interrupted the placid existence that was his and Ralph's world. Although I don't know what he had to complain about; I was, among other things, his personal maid.

Zamba took great joy in teasing me by waiting in ambush, then leaping out at me (Surprise!) threatening me with a massive coronary.

"Shoo–shoo–Go away! Or I shall bend your little whiskers and tie your precious pink tongue into knots," I would whisper. "Ralph, make Zamba stop it," I would yell, pinned to the wall in terror.

"Zam! Get in here, right this minute and leave her alone!"

("Okay, okay, okay, okay, okay. I can't ever have any fun. Jeezo, and she's such a perfect victim, too.")

After thumbing his nose at me–the neighborhood weakling–the lion would turn and prance off to lie by Ralph's side and peacefully watch TV. They had the couch; I got the chair.

Sometimes if I were to sit by Ralph, Zamba would rumble with audible grunts and raise his lip at me like a fierce, protecting dog, at which point Ralph would pop him on the nose and say, "Now stop that! Come on and sit down, Honey." And he'd pat the seat next to him fondly. Zamba would then make himself bigger with an extended stretch, pulling in a movement that would make him grow before our eyes. The continuous, unbroken length would scoot both Ralph and me off the divan to the floor below. Irritated, Ralph would stand before him, shaking his finger angrily as Zamba's eyes tightly squinted and his mouth formed the shape of a small *o*.

"That's not funny! Now, you get down right this minute." And, mumbling in his whiskers, Zamba would reluctantly obey. He'd lie pouting at Ralph's feet, watching us with injured dignity, quite irritated with me as victoriously I took my rightful seat upon the throne and, with unequivocal class, stuck out my tongue at him.–Well done!

We were on occasion no doubt in competition with one another over the master's affection. Sometimes Zam slept at the foot of our bed. More often he caused us great discomfort by sleeping between us, resting one great paw on Ralph's shoulder. I awoke many times with the feeling something was watching me, and I rolled over to meet the still gaze of a huge amber eye (which always gave me quite a start). Zamba slept with one eye partially open. (He also displayed a wide range of sound effects during his sleep: burp, whistle ... burp, whistle.) There were occasions when I would receive a huge lick on the face, causing my skin to feel as though it had just been surgically sanded. (It was not easy to get a night's sleep here.)

That old lion went everywhere with us–to the market, to the relatives, to the drive-in movies. He preferred Westerns, and often half way through the show, drifting into sleep, he'd jump and squirm and mumble, and I knew he was happily chasing a horse through tombstone territory, much to the dismay of a detached cowboy, who lay dazed by the side of the trail.

People certainly do ask questions when there's a lion in the back of your station wagon. Unfortunately they always ask the same questions: "What is it? Is it alive? How much does it eat? Aren't you afraid? Good heavens, are you aware there is a lion in your back seat?"

Finally, out of despair, I printed a little sign and placed it in the window of the car: It is a lion, he is alive, he consumes twenty pounds of food, you're damn right I'm afraid.

Ralph was much more direct with his answers than I. "This is a Texas cattle dog. Lucky Bugger! He was bred strictly to herd the longhorns. The hair around his head keeps the blazing sun from his body, much like the protection worn by the Arabs in the desert. You must remember we grow everything bigger in Texas," he denoted with a drawl and a level stare.

This was expounded with a great unquestioning authority, and people accepted his word as fact. "We call them Zambas," Ralph added. "If you should ever hear reference to a Zamba, you know now whence it came."

Lions are by nature extremely lazy, and one must coax them into movement. On arising in the morning I would numbly find my way to the kitchen and sometimes observe Zamba in one of his more regal positions–on his back with all four paws in the air, his enormous head tilted backwards, snoring and passing gas. When he was caught thus, it usually resulted in a loss of composure, and he would slip and slide under his great weight, trying to regain his dignity.

Often after such a duel with himself he would express a vocal tremor, at which my eyes would cross, and I would temporarily ripple from the deafening ear piercing. The windows were always steamed in our cabin from the intake and output of Zam's great nostrils. The acrid odor from his mouth was enough to enfeeble the mind, and he had the delightful habit of squirting all over you when you least expected it. Ralph kept insisting that this was his way of showing great affection, by marking me with his scent (a conspiracy, no doubt).

Zamba was prejudiced, a trait he picked up while working with the tribal people on the film in Africa. Two hundred curious Africans had charged the Landrover to see Zamba, terrifying him. He had never forgotten the incident. We had a trainer working for us who we all thought had black blood in him because Zam came completely unglued when this man came around. Ross had a good-natured face filled with appealing charm—and little else, for he was never burdened with an overabundance of brains. When I mentioned Ross's impediment to Ralph, he replied with a reassuring smile, "No, Darling, that's not an impediment; we all talk that way to the animals. It's our local jargon." But Ross spoke baby-talk most of the time and had the nauseating habit of displaying his muscles while preening himself before the mirror (which he often kissed, expressing some form of "wuvving" devotion). He exercised vigilantly a profound egotism, usually with simplicity and directness. He and his exalted senses would walk right up to a poor, unsuspecting female whom he and his hyper-thyroid condition had shown a vulture interest in and say with tears in his voice, "I'm so sorry I can't spend tonight with you. Here's my card. Call me tomorrow." And with that extraordinary approach, there were women who actually called him.

Ross was without question a part of the Africa USA scene. He was an industrious worker with an uncontrollable nervous energy. It was no secret that he fancied himself an actor and through Ralph saw an opportunity to get his enormous ego inside the studio walls.

Unfortunately, Ross never asked what he could do for the animals but rather what they could do for him. So, motivated by an ulterior motive, Ross directed a tireless effort toward his job as trainer and so assured himself work on the set, where he insisted on informing anyone whose attention he could hold of his vast experience as a performer. From his viewpoint, any production was in need of his talents. Occasionally he was put on as an extra to humor him in his bold exploits. One had to give him credit. For a man of few words he had come a long way on boyish charm.

Ross had one nasty little habit that was never cured. He was a thief, who stole from anyone at every opportunity not only material things but things emotional as well. Although he loved Ralph and was a devoted friend on a one-to-one basis, there wasn't a thing Ralph had that Ross didn't try at some time to take. And when a career as a Hollywood star never materialized, he struggled with a sickening effort to assume Ralph's identity as his own. But even when Ross was his most nauseating self, one couldn't stay mad or hold a grudge against him for long, for Ross was blessed with charisma.

I had never seen Zamba in action. To me he was now just a big, old loveable mush whom I too often took for granted. One morning Ralph had gone on an early studio call. Ross came to the door to try his luck with me while the boss was away. It wasn't that I was a femme fatale, mind you; it was more that Ross wasn't particular. He had recently been divorced, and women weren't safe within ten miles of him. Not that they were when he was married. He'd been married so often none of us could keep up with the who's who. Anyway, there I stood in my feminine Brunhilda best, wearing Mukluks and Ralph's bathrobe, looking as though someone had just dug me up. I said to Ross, "Thank you for the offer, but I am not interested, now or anytime for that matter, and if you know what's good for you, you'll get back to work in a hurry!" And I slammed the door in his face.

I went back into the kitchen to tidy up before going to work, and as I stood at a sizzling sink of hot water filled with dirty dishes, to my astonishment an arm draped itself around my waist. There stood Ross, who longingly said, "Woo didwen't weally mean dat, did woo? Woo naughty little fing."

I found that last part most distressing. "Yes, I meant it. Out! Get out of here, you–you peasant! Now!" I screamed as I stepped on his oversized foot, "Get out of my house!" At this point Zamba, seeing Ross in the cabin, went completely berserk and turned into a ferocious rage-filled 529-pound African lion intent on the kill. He flew at the flimsy chain-link fence dividing his room from the kitchen and let out a blood-curdling "Roaaaaarh!" The divider crashed to the ground, and Zam was after Ross with glazed, abnormally wide eyes. I was absolutely mortified, not to mention turned to stone. (Good grief! A lion actually lived in there!) Ross moved with the speed of all winged things, making it to the door just in time, and with a fierce thrust he slammed it shut in Zamba's face, giving him an awful nosebleed and making him even more angry. Meanwhile I was left alone, helpless with this "killer," who was very discontent at having been outsmarted by his prey. He paced back and forth in front of the big bay window in the living room and for some blessed reason didn't go through it. Having not a second thought, I went right out the open window above the steaming sink in the kitchen. There was *no way* I was staying there. I'd rather have fought off Ross, who was long gone. Somehow I coaxed Zam from the house into his outside pen; then I ran inside, locked off the kitty door, and fell, in a feverish sweat, trembling against the wall. I repaired everything properly and never told Ralph that I was now aware Bert Lahr did not in fact inhabit that great fuzzy suit. Still, it was nice having a big, strong, healthy friend.

4

During some of my first visits to Nature's Haven, while Ralph and I were still courting, I had learned from him valuable lessons about the animals on the ranch.

One hundred yards behind the corrals stood a huge flat rock, twenty or thirty feet high and equally as wide. After one of our hikes, we would recline there by instinct. Perched high upon that rise, I felt a diminishing of my fears, as Ralph began to teach me. And with each lesson he usually told me a story. My tour of educational animal behaviorism had begun.

Some of the classes he held for my benefit were a little startling, as was the one regarding the dangers of my being around certain adult male animals. (It's a jungle out there, my mother had warned me. Little did she know.)

"Toni."

Ralph's voice had a concerned uneasiness about it.

"Uh huh," I turned over on my side, propped my head on my right hand, braced my elbow, and watched him thinking out loud.

He took a deep breath and said, "I am going to explain something that might be a topic of delicate conversation, but it's a subject we need to share together for both our sakes. All right?"

"All right. Sure." I never knew what to expect from him.

"Have you noticed the chart hanging in the tack room?" he asked.

"The one with all the girls' names listed alphabetically on it and the *big bold* letters checking off the days of each week of the current month? No, I didn't notice it, and I don't think I want to know why it's there, but I have a feeling you're going to tell me anyway."

"That list of women's names happens to represent an element of extreme danger. For years now I have had two such charts, one that hangs in that place of secluded convenience and one that accompanies me to the studio. It states the name and dates of the menstrual cycle of each female employee—and you know, that's a considerable number—and every frequent female visitor to the ranch."

I wasn't sure what I was supposed to say to that unusual confession. So I kept quiet.

"Some adult animals, primarily the primates, felines, and bears," he began, "have proven to be predictably dangerous during the female's time of the month. Whereas the human male is generally disinterested by menstruation, the male animal, to the contrary, is stimulated not only during the flow but several days before and after. Since the animals have intercourse during this time with their own kind, they quite naturally relate to the female homosapien in an affectionate way." (What a terrifying thought!)

"It goes without saying that this is a sensitive subject, but I have to explain all the hazardous elements involved, or it could cost someone a life. Are you still with me?"

"Of course." (One potato, two potato, three potato, four. I always resort to this verse from my childhood when I am confronted with stress.)

"If," he continued, "I am aware ahead of time, I can regulate the situation. And if you are going to be involved with the animals, your odor can be controlled by internal cleanliness as well as camouflaged with other smells." (What would my mother think if she were here?)

"You see, sex is one of the animal's strongest motivations, and possessiveness is a completely natural trait. This is where the danger comes in for the woman. The animal's possessive attitude toward you, not the sex act, is what could expose you to harm. A stallion breeding a mare holds her by the back of her neck with his teeth, and a cat holds his mate with his claws and teeth. If she moves, he controls her movement by physical domination. It's not so much a matter of the animal intentionally trying to hurt you, but his amorous intentions could get out of control in his attempt to relieve himself in any way he possibly can."

"You mean!"

"Yes," he said, "by masturbation," using naturally the word I was having difficulty saying.

But he went on to explain an even greater hazard.

"The danger to the male trainer involved is one of being caught in a sexual triangle. It then becomes a case of male ego pitted against male ego: She's mine; no, she's mine. If a girl is a victim of the social dark ages and her shyness causes her embarrassment, she is unable to admit to her situation. It's unfortunate that this condition is still discussed in a hush-hush manner. It's a perfectly natural part of a woman's life. Do you understand?"

"Certainly." (Understand? I was dying.)

"Let me take this one step further and give you a bit of technical knowledge. I establish respect when I evolve a pattern of taking an animal from his quarters. First, I determine the mood he or she is in, and if it appears to be a good mood, I either take the animal out for the pure pleasure of it so that we might have fun (in which case the control is less), or I take him out with a firm voice and with complete control as though we were going to work–in which case the teacher-pupil relationship exists. In this respect, early control is established and qualifies me as the boss. So work and pleasure each has its time.

"If I have oneupmanship when a dangerous situation arises, I can control that problem to a certain extent. But if the difficulty comes during a play session I have lost my regulating influence. Does that make sense to you?"

I glanced over at the sleeping lion who was snoring and casually breaking wind. He certainly didn't look like a sex offender.

"It means," I responded, "will you sign in, please."

"I'm afraid so," he commiserated. "Just the moment you regain consciousness."

There were of course a few other milestones I would have to overcome, but by far the most difficult for me was entering my name on that chart that hung like a scarlet letter in the tack room for all to see.

–Don't tell me about suffering and personal sacrifice for the man you love.

Only one week after our intimate discussion Ralph was interviewing a group of undersized women, one of whom might be chosen to double for the little girl (Pamela Franklin) who was to play opposite Zamba in the movie *The Lion.* (I had considered chopping my legs off at the knee.)

Instead I sent one of my short girlfriends out to audition for the part. The motive? If she were chosen, I would know his every move while he was away. It's a dog-eat-dog world.

On that bright Sunday afternoon, with his usual scientific directness he began to deliver the same sensitive speech I had recently heard.

"So you see ladies," he concluded, "As long as we are made aware that you are in or near your period, it will not interfere with whether you qualify for the job."

There was an embarrassed hush from the seven women. Not one of them raised her hand. (Death before dishonor.) And so the men were forced to accept their word. People in Hollywood have a strange habit of never admitting to anything during an interview. If they think their chance of getting the job is in jeopardy, they will usually do whatever it takes to keep themselves in the running.

Stuart Raffill was Ralph's back-up man. He had been his apprentice for nearly a year and in years to come went on to become an extremely good producer and director of animal films in Hollywood. Having read of Ralph's work, he had emigrated from England at the age of nineteen. Unannounced, he appeared one summer day on the corner of Sunset and Vine, holding a satchel and possessing little else than class and festive charisma. Six feet seven inches, slim and on the verge of blooming good looks, Stuart, like a meteorite, descended upon the ranch. He had a certain strength born from a kind of indefinable desperation. On impulse he had left everything behind him and with full sails had set out to take on the world in waves of perennial protraction. No obstacle stood in his way other than possibly his elongated self. The wind had come with Stuart, and he reacted from inside a tornado. All things fell or were uncontrollably knocked down at the sight of him.

Ralph and Stuart began to introduce formally, one at a time, each of the girls to the five-year-old male African lion–a fine lion, one who had never acted offensively. There would be days of interviews, after which Ralph was to send the three girls he felt best qualified for the job over to the studio, where they would make the final decision.

The last of the girls–my friend, Miss Righteous Indignation–was about to meet the lion. She came forward to satisfy a long dreamed-of thing–to touch a big cat. Without warning she sat down upon the ground. The superb creature flopped down beside her and laid his head in her lap as she began stroking and cooing and brushing his mane with her fingers as a small child might comb the curls of her doll.

"Miss, please get up off the ground; that is not a good position for you to be in," Ralph insisted.

"But he loves me," she said indignantly.

"I asked you to get up!" he ordered.

With a slight tug of the leash Ralph suggested to the lion that it was time to go. He responded with a low growl, and, though sitting up, stayed where he was.

"See?" she echoed. "He doesn't want to leave me."

Ralph tugged a second time. A hollow rumble began as the cat's tail began to lash about, and he barked an angry, defiant response. Then the outraged lion, with pinned-back ears and blazing eyes, jumped and attacked Ralph, biting not only into his shoulder but puncturing a hole in his left arm. While Ralph lay bloody and twisting upon the ground, the cat strutted back to my terror-stricken friend, who–fortunately for her–had been too petrified to move. Before Ralph could get to his feet, again the lion charged and slashed.

When animals have these flurries among themselves, they heartily go at it, but rarely connect. When the onslaught is directed at a person, however, that person cannot physically sustain the weight, as another lion could. The animal does not intend to kill, but simply to get the obstacle out of the way so that he may continue, undisturbed by competitive intrusions, with his amorous adventure.

At the time of attack, rarely does a backup man get to the scene before the animal's first charge. But hopefully he can distract the animal's attention to avoid a second. Had Stuart been holding the leash, the assault would have been directed at him because he would have been in control. Others, though in danger, are generally exonerated from immediate rage unless too much opposition makes the animal react. To intercede is the backup man's number one function. The guarantee that one man will risk his life for another is never assumed–though most certainly hoped for.

Somehow Stu had managed to tie the lion's leash to a nearby stake. The cat was not at all happy about having been separated from the girl of his dreams. My friend, in the meantime, before bursting into tears, confessed in a fit of hysteria that she wouldn't have been able to pet the lion if she had admitted to her condition and that she hadn't believed what they were saying about such a thing.

I wasn't to find out for a long time that it was my friend who had caused the attack that day. She never mentioned it to me, and only once did Ralph bring it up, years later.

5

Our ranch, Africa USA (with the move, so had the name been changed), lay beneath the torrid sun in Soledad Canyon, forty-five miles from Los Angeles at an altitude of 1,400 feet and just ten miles from where the old Nature's Haven was carelessly being chopped in half. Rough-cut though Soledad Canyon was, for the time being it was the only entrance into the Palmdale-Lancaster basin. The rugged, winding road supported an extraordinary amount of traffic as well as more than its share of accidents, the pieces of which we took to the hospital several times a year.

The property had been sold to us at a reduced price, since the previous owner had been shot and murdered in a snack bar by an inmate from the nearby detention camp–none of which did my sense of well-being any good. Across the road from our front gate was the entrance to Indian Canyon. The foot of the mountain, for nearly two miles, was covered in a thick tapestry of brush and heavily inundated with a rich dimension of trees, some deciduous, others evergreen, the colors of which stood out against the irregular environment.

To the back of our ranch, just across the Southern Pacific railroad track (better a railroad than a freeway) a succession of small burnt hills, washed dry and rendered brittle by the parched Santa Ana winds, took fire at least once each year. Here Boulder Mountain soars above the surrounding emptiness. Birds of prey nest within its pock-marked surface, and their calcium deposits dribble down its sides, giving the ominous rock a cadaverous appearance.

The ranch had been a summer resort for campers and picnickers. There were snack bars and dance floors, rows of picnic tables, enormous brick barbeques, and green, spacious lawns edged in log railing. Cottonwood trees and live oaks formed a sunscreen above our heads, and through the lush 600 acres sparkled a stream whose tributary filled each of the two large lakes, then weaved its way to the sea. At night the animals came out of the hills to drink from our stream, and because of their need the ranch was not fenced in on that far side.

At the top of a small man-made hill we built the food preparation room. Surrounding it was a large circle of cages in which the cats and bears lived. Below were the hoofstock pens and the large wooden building that had formerly been used as a mechanic's garage, now reinforced and converted into an elephant barn. Trailers had been parked in the upper lot near the snack bar, liberated from the homey life and transformed into offices. On the lakes all the birds were turned loose to live freely among the lovely pampas grass that gave the ranch a quality of great beauty, rather like an estate whose grand house was yet to be built. And although the work was hard and the going sometimes terribly rough, Africa USA was a place that took hold of us all, and we loved her with a protective, ardent devotion, working with boundless enthusiasm, accepting without resistance all she could hand us.

4
The Novice

Cries still are heard in secret nooks
Til hushed with gag or slit or thud,
And hideous dens whereon none looks
Are blotched with needless blood.
But here, in battles, patient, slow
Much has been won—more, maybe, than we know—
And on we labor stressful. "Ailinon!"
A mighty voice calls "And may the good prevail!"
And blessed are the merciful!
Calls yet a mightier one.

—Thomas Hardy

1

Until now I had been an observer at the compound, and, quite frankly, the participant sector hadn't occurred to me. But the time had come. I was gradually introduced to every animal, and with daily religion as well as wavering frustration, I began to make it work for me. There were no books, no records of traditional, ancestral past, no social logs to help me. Besides, my thoughts went about as deep as "How would I look on that horse?" or "Wouldn't I look swell in that car?"—"Do you like my hair when it's up or when it's down?" "How would I look as a nun?" So I was taught to read the animal's feelings on their terms, and only experience coupled with Ralph's guidance could prepare me for communicating with them. As the soft sounds and the sweet smell of cubs began to lift me, the wooden floors of my home were suddenly scattered with scenes of wild-life—social and unsocial. Together the cubs and I grew up. For a while, though, I felt as though I was the mother of the world. I lived with baby bottles, formulas, feeding schedules, diapers (for the apes, you know), vitamins, paper training, "Good boy, bad boy," tummy aches, and all-night fevers. My social life suffered considerably, but after a year or two I could sit and talk lion or tiger or wolf or bear with the best of them.

I now stood before the puffing tiger's cage as a Christian must have stood in the Coliseum. Ralph opened the chain-link gate and clipped a leash around Patty's neck. The two of them, in a boundless pace, came my way.

"Wait! Wait! I wasn't made for combat!" I whined. Only yesterday I had fancied myself a woman who would follow her man into battle, to fight beside him in a heroine sort of way. But in the face of action, I was riveted to the ground. The enemy was advancing, and I had completely forgotten the campaign I had planned so diligently for this moment. In my cowardice I began to retreat.

"Toni, stand still!" Ralph dictated.

"I'll force myself." I grimaced under my breath. Being review conscious has its definite disadvantages.

"Now kneel down just a bit, so you're on the same eye level with her and puff back when she puffs at you."

The tiger puffed at me, I puffed at her, she puffed back and so on.

"Oh, no, you're too kind." I blushed amidst our puffing contest, as the tiger put a huge paw on my knee. "Do you mind?" I said with controlled alarm, and she puffed a spray directly into my face. "I do wish you would learn how to do that with less showering effects. You know, it's very hard to take you seriously when you are making quacky duck sounds." ("Better than breathing fire in your face," a little voice in the back of my head reminded me.)

"You're doing great, Toni, just great!" Ralph acknowledged. "You can stand up now—just keep petting her on the sides and talking to her. Occasionally let her lick your arm, keeping your fingers closed so she can't get in a little bite."

(And I was doing so well until he said that. Back to the bench.)

"A little bite? As in missing-our-finger time? What a sobering thought. Well, we all have our little headaches."

He smiled. I did not find it amusing. "Patty wouldn't hurt you for the world," he reassured me as we began our walk around the ranch. Your fingers are a vulnerable part of your anatomy. They have a tendency to get in the way. I just want to make you aware of their susceptibility to injury."

(The things one has to go through for a little creature comfort here!)

"Here, I'm going to pass you the leash, and you walk Patty for a while."

"All right, if you insist, come along my darling." And I began with an overreaction of tiger puffing.

"Hold the ring with one firm hand and grab the center of the chain with the other. If you give her too much slack, she has room enough to turn and pounce on you."

"*What*!" I cried.

"It's listen-and-learn time," Ralph philosophized. "Listen and learn! First, a lengthy observation of an animal's idiosyncrasies and its attitude is necessary before I can train you or any animal. (I wasn't sure I was ready to be referred to as an "animal" yet.)

"Emotions are a strength that link all living things. Through emotions we achieve our control. They allow for the interaction between the species. Love, hate, fear, anxiety, isolation, companionship, security, insecurity, and stress are a few traits that make up the animal condition. In the development of my theory I have concentrated on the evolution of mutual understanding. You see, animals share, in varying degrees, psychological experiences that closely parallel many of our own." (I'd be willing to bet they wouldn't die of fright!)

"To establish a natural basis, you may often find yourself working from spontaneous impulse. By keeping one sensitive step ahead of your animal, you insure safety. By that I mean you *must* see the world through the animal's eyes, always working on the defensive, prepared for the unexpected that might startle him. (And you think *you* have problems!)

"When you're out walking, as we're doing now, if you are in tune with your surroundings, you are in control of any situation that might arise. For instance, if a jackrabbit were standing in the trail just ahead of us, you would instantly divert her attention to something else, perhaps by turning her around. An alert mind is a marvelous thing when you are confronted with a dangerous situation."

The three of us were now jogging around the lake at Beaver Dam. I experienced the elation of a team feeling. Ralph bent down, never skipping a beat, and whispered something in Patty's ear.

"What did you say?" I panted.

"I put in a good word for you," he inhaled.

"Oh, capital! I do appreciate it," I gasped.

"It's during these quiet times of ambience," he began as though catching his second wind, "that you will really begin, with full concentration, to develop and establish a kinship.

"I have always responded to my feelings about an animal's nature, exploring and drawing upon a single impulse or a slight inflection of its multifaceted personality, which conflicts and varies as much in an animal as it does in a man. And that brings up another point we should discuss. The *balance* of the animal's and the trainer's personalities is imperative.

"For instance, I, being a very calm person, will equalize and counterbalance an exceptionally nervous animal. These first stages of establishing a man-animal relationship of interdependency and trust are based on mutual respect, positive motivation, and unquestionable acceptance of one another. You must distinguish each animal in its own social sphere.

"I use a combination of methods to adjust people and animals to a state of good mental health. Through affection training I can relax their anxieties and replace their insecurities with a strong feeling of confidence when I am around. But in order to develop personalities and reinforce natural behaviors, I must go through a period of trial and error and often fatiguing mental observation," he concluded.

Patty had been a three-year-old Bengal tigress when she was donated to Ralph. She was already a young adult with wild habits, tribal mannerisms, and an established set of wild values.

"*Training*!" Ralph exclaimed, causing me to jump. "Now there's a word that has been misconstrued. Training animals is unfortunately associated with 'sit up and beg', or 'lie down and roll over.' These are 'tricks' that many people–because of circus conditioning–expect from my (*our,* he corrected himself, and I beamed) animals. These tricks are a deception I do not cultivate unless there is a very good reason for it, such as to prove to a circus trainer that they do not have to employ fear training methods to achieve their tricks–or if a film specifically calls for them. When I refer to training, I simply mean *teaching.* Our animals are educated!"

Patty looked lovingly at Ralph and puffed with an acknowledging roll of noteworthy whispers.

"There was a time when I didn't have the patience to deal with *wild* full-grown animals," Ralph revealed. "Patty was given to me when I felt the need to advance my technique–to bring adult wild animals (not just cubs) out of their terrifying anxiety and mistrust into the comfort of their surroundings. If I could do that, I felt, I would have really succeeded in breaking through the conscious barriers between man and animal."

"Ralph?" I panted.

"Yes?"

"Do you think we could take just a little break? We've been around the lake ten times." I don't know which of us crashed first, Patty or me. Patty's tongue was hanging out, and if I'd had a brown bag I would have placed it over my mouth and nose.

"How about a swim?" I invited.

"No thanks. You two go ahead." He grinned. "I'll watch."

I looked at Patty, she looked at me, we looked at Ralph–who was looking at the sky. He paused, stretched his arms out, and exploded, *"Isn't it great to be alive!"* Patty was used to these glorious outbursts. She just sat there. But Ralph always succeeds in scaring the feline right out of me.

2

I wanted so much to be accepted and to belong. Other than Ralph, Bert, and Gloria, there were those who, even after a year, considered me an outsider. Often I had to battle against hostile odds even to qualify. I experienced a brief cultural shock from the crash exposure. I was a long time proving myself. Only when the cubs I had been raising began to outgrow the house did anyone discover my singular contribution to early identity.

A large stone trestle stood in the park like a monument to the rerouted railroad. I thought that to this sturdy structure we should attach a building where I could properly raise the young animals, which were "popping out" this spring at such a rate Gloria and I were maternally overcome. Gloria was the dietician. I thought her very attractive. She had the look of one who had soared among social circles. She was Laurel's mother, and she lived agreeably with Bert. Although there was a ten-year difference in their ages, Gloria never showed the decline of her years, having still the elasticity of youth.

When she and her daughter had met Ralph at the Armenta Ranch, it was during the time she sought adventure so often yearned for by the middle-aged. At that time she discarded her life of married security and never looked back. She was always attracted to younger men, who constantly interfered with her "affairs." Gloria's lack of interest in anyone past forty seemed centered around their lack of mystery. To her, over forty represented a finished novel, men who were locked into their ideas and set in their ways, stable, stationary, and stagnant.

We got along very well. I treated her as a mother would treat her child. She would bring her cubs over to play with my cubs. Those were times of great happiness for both of us, surrounded by our wild babies and she passing on to me information like a retired nurse from a child-care nursery school. We gossiped across a clothesline of drying primate apparel about every piece of news regarding the ranch. From Gloria I learned all she had to teach about the rearing of young animals. And together we talked Ralph into building the nursery.

If there ever was a heart or a pulse of a place it was that nursery. Within its spacious walls young animals were nourished and reared, set apart for development and growth. It was a gratifying thing to be surrounded by so much happiness. And my little flock followed me everywhere.

The building was quite large, far more spacious than necessary, so it was divided by a wall and cut in half with a door in the middle. I now had to walk down a row of closely spaced large primates, who reached out through warm, barred indoor winter quarters and attempted in every way possible to siphon me in, at times retaining handfuls of my hair or bits of my clothing. I often felt as if I were the only sane soul in an insane asylum.

In the nursery was a preparation room of sinks and refrigerators, stoves, and storage shelves. And to the right was a huge, glassed-in, thermostatically controlled room containing gnarled driftwood; entwined around it were Loch Ness and her friends. Ralph had reluctantly moved the snakes from "my" house to this new apartment with a view. Molly, the African grey parrot whose entire repertoire consisted of an eyebrow-raising recital of filthy (not just dirty) four-letter words, perched freely in the far corner where she could accost anyone entering. She was rather like the black sheep of our bird family, and I really should have kept her in a place of isolation; her vocabulary would make a truck driver blush. The building was again divided, this time by a six-foot chain-link fence with a center gate. Beyond the gate was a long, wide, dirt runway, and on either side were five little rooms with double dutch doors and wooden floors, each with its own little shelf for an animal to lie upon, steam-cleaned tires for teething, and sawdust bedding for rolling in. This was where the babies (having reached a certain age) napped and spent the night. During the day all twenty of the original brood attacked and stalked and climbed trees and pounced on one another from the fenced-in enclosure attached outside the nursery building.

When an animal was born on the ranch, its arrival was celebrated for months. And while the baby remained with its parents, it maintained a high self-esteem. You could almost see the spirit within the body rise and become bigger than the animals themselves. The mothers who accepted their cubs had a sense of dignity, and they glowed with pride when we came to visit. Because our animals were so close to us and since no dividing barriers isolated them from us, they would often allow us in their homes to play with their cubs. They seemed to draw great pleasure from sharing their deepest feelings, and more than once we saw maternal lights in those gentle eyes.

When the cubs were on the threshold of achieving independence, they took up residence in my nursery, and here they began their primary education. Before I learned to speak animal, I had a curious relationship with them, for even the tiniest among them would become increasingly frustrated with my ignorance of their language. Sometimes when I made the wrong sounds, passing down to them some incommunicable noise, they would look up at me as though I were possibly the dumbest creature on earth. At other times, mispronouncing a sound, I would get a startled reaction from my furry little teachers, as though I had just said something foul and distasteful. "What did I say? What did I say?" I asked, experiencing a spontaneous remission to their benumbed reaction. I could only assume that I had referred to a member of their family with limited perception.

I began with a tape recorder to transfer their sounds onto paper, recording them in the same fashion a composer would write music. They found me terribly amusing in the classroom because I was continually vibrating with the use of my new vocal achievement, at the same time trying to define a cultural system of communication between us. I sounded as though I were making some desperate attempt to speak with a violin in a high register.

It was much easier talking with the primates, who were able to relate with ease to my tongue. I had but to establish a symbolic order of hand language that Ralph had designed for me—in a simple fashion at first so that the primates and I were in accord. That they learned and responded to my hand language faster than I did to theirs made me feel somewhat inferior. It was fascinating and completely humiliating to give an elephant or a chimpanzee a list of things to do and to have them not only understand but actually follow through on a level of excellence. But when one of them tried to get me to do something, I stumbled around like a great ignoramus, trying to interpret its meaning. That these animals could understand me and I could not comprehend them was highly embarrassing. They found it very strange indeed that I was not able to follow their train of thought for any length of time.

The attitude of the public about wild animals was then one of two things: that animals were savage and treacherous, to be feared—or that they had a handicap, attesting to the emotional gymnastics of people who sat in judgment of a subject they knew nothing about. At Africa USA we were developing a support group, as we saw it, for lives with special needs.

It seemed to me to be the beginning of a distinct era of sensitive understanding. We were all absorbed in and preoccupied with contemporary animal knowledge and the advancement of our individual discoveries.

The wild animals had about them a magnificent spirit. They carried their heads high and walked with great self-respect, as if they were deeply satisfied at having been born as they were. I always felt in the company of superior beings when we were together. Even when they were tiny things, I never felt equal to such a self-image. Although we all became the best of friends, I could not help, at times, feeling I was being patronized–that perhaps I had been placed here especially to serve them.

The animals and I looked good together, and we were photographed and written about for many years. But if I had not had a life among them, I doubt that any newsman would have had much interest in just another beige person.

Most of the animals enjoyed showing off, particularly those with an accelerated intelligence. They displayed humor at every opportunity and made one another laugh aloud. And they got great satisfaction from making a fool out of me. I was never to be feared for any length of time, for they had my number, right from the start. They treated me as a retarded member of their kind and were sympathetic in their dealings with me.

Ralph, on the other hand, was one of them and held a position of the highest order. He understood them completely; there was never a lack of communication between them. But I was forever doing my homework,

Toni: "In the company of superior beings."

Toni and Yang–Africa USA In winter.

Toni hand-feeding her cheetahs.

Surrounded by fur-covered Lilliputians.

trying to keep up with my own natural instinct. Sometimes it hurt my feelings when Ralph would come into the nursery and suddenly divert the attention that had been centering around me. King of the Animals, their Lord, was present. He sat down on the sawdust-covered floor, and they gathered around the one who held dominion over them. I was their mother, yet they left me for him. And after he had said some esoteric thing (unknown to me but perfectly clear to them), after he had made a brief inspection, he would leave. Then I was suddenly set upon, brought down laughing by my clowning bunch, and tickled into submission. It was frustrating–this power he had over them that I was never able to achieve.

My twenty cubs and pups were, within various days, six months old at the time their expensive play pen was finished. The greater part of our days was spent in happy heroics on the parklike lawn in front of the nursery. In my brood were three tigers: Sultan, Penda, and Sarang; two mountain lions: Sierra and Sequoia; two African leopards: Yen and Yang; two Indian leopards: Riff and Raff; three lions: Ivan, Gypsy, and Zamba

III; and one cheetah: Tara. There were two orangutans: Hannibal and Genghis Kahn; three jaguars: Zoiliva, Hodari, and Changa; one wolf: Little Wolf; one sloth bear: Dinky; one spotted hyena: Kenya; and one American black bear: Mike. It was a matriarchal society over which I ruled with a rolled-up newspaper.

During this first voyage I was surrounded by tiny, fur-covered Lilliputians, who, upon finding me marooned and alone, tried their best to take me prisoner in this land of twelve-inch wildlife. Taking themselves quite seriously, several of them, with little scoots and squats, hid behind native trees. It was apparent I was under ambush, and, with a light rustle of grass, I was set upon by the small but ferocious assailants. With a demonstrative roar I surprised them and sent them scattering head over heels in all directions. Scarcely had I made good my flight when stripes and spots and all kinds of furry things hurled themselves on me from out of the trees, showing no mercy. It was a bountiful profusion of cubs that assailed me that day, sorely knocking me about and attempting to drag home my bruised bones. I was, with the lot upon me, compelled to surrender myself; for the disposition of this tiny militia was fierce as they pinched and kneed, shredding my clothing and pouncing upon my body. And when I fell among the rascals, locked in combat, I redeemed my life by paying the specified sum of twenty bowls of milk and eggs. They set upon this ransom with ravenous appetites, and with this diversion I regained some ground. On my hands and knees, with my body close to the soil, I crept toward the shore, from where I had made my way. But quick as they were, and seeing that I intended to make my escape, they bolted full-force my way. I stood my tall size, and, running for dear life, followed by hairy things, I fled toward the nursery. Through the first door, past the fence gate, and into the cage. They were on my heels as I plowed back through them and hastily shut the gate. I clinched my first victory, threw my arms into the air, and shouted, "She lives! Daniela lives!"

3

There is something self-incriminating about observing hanging flesh, and I've never been able to justify life's macabre banquet of the living feasting off the dead. I have not always been a vegetarian, but I most surely have always agreed with Thoreau's opinion on the subject: "I have no doubt that as a part of the destiny of the human race in its gradual improvement we shall stop eating animals as surely as the savage tribes have stopped eating each other."

Since we are all terminable–death lives within each of us–it is sometimes not difficult to treat final departure as perfectly normal (when we are not personally involved). But there was never stimulus for either conversation or thought about such things while I worked with death. Farmers from far and near brought their deceased horses and cows to us as food for the carnivores. As long as they had died an undiseased death, we accepted them thankfully, for they saved tremendously on our food bill.

Ralph had set a hoist and freezer back by Beaver Dam, and for about six months there was no one to do the butchering other than my involuntary self. My first encounter with bloodshed caused me to throw up. That awful cliché, "Someone has to do it," was my only inducement. I gazed for a long time at the hanging horse carcass, trying to block out the thought of a personality having inhabited the suspended form. It was a hot, blistering, saltless day, and the longer I put off the sickening chore, the bigger the animal seemed to grow. I tried to conquer as many excuses as I could for not beginning. "The cats are too fat" (a small solace) was the only legitimate defense I could muster. I looked up for strength and instead found vultures eagerly circling and drooling overhead.

"Give me strength," I yelled as the enormous meat cleaver and I ran toward the hanging thing. I sank the cleaver in with the apologies of the living, and the carcass pardoned the injustice by exploding its entire contents all over me. I had unknowingly allowed it to bloat.

The idea of what had just happened to me was so inconceivable that I couldn't catch my breath. I stood motionless, gasping, draped in coiled intestines. After that disgusting initiation, my career was launched. But there must have been some mistake; this was not the club I had in mind joining.

5
Black Pegasus

"Hast thou named all the birds
without a gun?"

—Ralph Waldo Emerson

I remember King through a veil of perfection. There was never an unhappy moment between us. This is his story, which begins with the story of another horse.

We had a champion Appaloosa stallion named Dakota Duke. He was a very famous horse and during his show career had decorated the covers of many equestrian magazines. Duke's reputation spread from coast to coast, and people came from far and near to breed their mares to him. In the years of Duke's retirement from the show ring we received more requests for his stud services than he could handle. With the highest stud fee around, we had a very compatible partnership.

One day a dreadful old cowboy drove up and parked his dilapidated truck in front of the bamboo gates. In a crude and arrogant manner he announced to the guard that he was here to discuss the terms of "breedin' " his mare to Duke. "So get them Helfers down here so I can get this show on the road."

He was the kind of person you didn't want to touch, much less shake hands with, for fear some of it might come off on you. He had two black teeth in front of his cold-sore split lips. Apparently he kept these two just so he could make people like me nauseated by spitting tobacco through them. His clothes reeked of perspiration. His cowboy hat was burdened with grease and sweat, and when he removed it to say "Howdy, Ma'am" to me, I noticed a permanent dent circling his pear-shaped head. There was enough dirt under his nails "to grow potaters in." Of one thing there was no doubt: he was a mess!

"I want a colt from that there stud-a-yers," he eloquently divulged.

The picture entered my mind of a very large shot of penicillin awaiting his unfortunate dobbin. As it happened, the man had no money. He suggested making payments of no less than ten dollars a month. "But you would have your colt before we would have our money," I replied. He spat. "Patue." I turned green.

Ralph, with great patience, stressed that our books were not set up to deal with such a transaction. we were very sorry, but unless the full amount was paid up-front, the arrangement just wouldn't be possible.

The man stamped his dirty, mud-caked boot on the clean earth and spat in his hands, then rubbed them together. This time I went weak, and I had to turn away. "Okay, I'll tell ya what I'll do." (I could see a used horse salesman about to take over.) "I've got me this big black stud at my place. He ain't no good to me. Can't ride him," he mumbled, then caught himself. "But that don't mean you can't. I'll give him to ya fer the breedin'." He beamed under the delusion that that was some kind of salesmanship.

We smiled and much to his surprise said, "No thank you." As we turned and began to walk up the sandy hill towards our office, he shouted, "I'm gonna give him to the glue factory if you don't take him."

We continued walking.

"Yup, that's what I'm gonna do. I'll let 'em kill the no-good—"

Ralph glanced at me with one of those looks that said, "That's all we need, another mouth to feed," took a profound breath, shook his head, turned and capitulated. "All right, bring him over on Tuesday morning. We'll work something out."

"You got it!" the man yelled as he jumped into his truck before we had a chance to change our minds. He drove down the old canyon with the thrashing and popping of his deficient engine interrupting the silence there before him.

When Tuesday morning came, I had forgotten about the incident. I was driving our Ford tractor and ploughing under a new section of field to add to the existing veldt.

We had enclosed twenty-two acres of shrubs, trees, and streams. More than fifty animals roamed this giant sandbox. As my little tractor and I drove in a cloud of dust, we were frantically pursued by Bullfrog, our Indian water buffalo.

Since his birth he has had an aversion to engines and persistently tries to kill every vehicle he ever comes in contact with. The challenge of dodging him made for more fun than I got work done.

I was so enveloped in dust I didn't see the rusty old truck and trailer pull into the ranch. The voice came on my walkie talkie, "Your new horse is here, Mrs. Helfer." I couldn't imagine what he was talking about. Then I saw the waving, spitting, greasy man.

"Get Mr. Helfer," I called. I shuddered to think what he'd brought.

Ralph was pensively indifferent; his interruption had come at an objectionable moment, and when he came down the hill toward the homemade vehicle resembling an open tin can, he was in no mood for the cowboy's shenanigans.

The trailer was so small that the tall black horse occupying it was literally wedged in. It took a team of us to get the poor creature out. His head hung low as he inched his way backward, and when he was finally released, he stiffened and crashed to the ground, where he remained still except for the occasional thrashing of his wobbling head.

"Is he dying?" I asked with grim alarm.

"Well, if he isn't, it's not because that broncobuster hasn't done all he could to kill him," came Ralph's expressionless reply.

We were all so taken aback by the pathetic sight before us that few noticed as the cowboy crept past in an effort to escape. Outnumbered by a maddened crowd, he was more than fortunate to get away with limbs intact. Had he stayed, he would not have been standing long.

"Toni, run to the clinic and get Marty as fast as you can! George, you hold this guy's head so he can't injure himself." Everyone began functioning regimentally under Ralph's barked commands as we had all trained to do in an emergency. Suddenly it seemed to matter a great deal that this horse live.

The clinic was quite sophisticated by any medical standards and, with our work becoming daily more demanding, Ralph had hired a young veterinarian just graduated from U.C. Davis. Marty Dinnes, in the short time he had been associated with us, had developed a great passion for his work, and in these desirable training grounds he would have a future to elaborate on. Marty was tall, slender, and informal with that coppery, chili-pepper color only red-haired people seem to have. He looked much like Michael Caine. Besides a good mind, his outstanding quality was a vivacious, bubbly personality that made him hard not to love. Marty's heart, though never very strong on its own, was dedicated to an exhausting career, to which he applied himself with the devotion of a well-schooled professional. During the many years we were all together we shared sufficient tears to last a lifetime and enough joy to fill me with stabs of nostalgia whenever I'm alone.

I helped load Marty's medical case, and we raced with frightening speed to the spot where the black horse lay motionless.

"Poor old fella," Marty said as he patted the animal. "There's a good boy." He listened to his heart with the stethoscope, inserted a rectal thermometer, checked the walled eyes, selected a syringe, a long needle, and a bottle, and injected the contents directly into the horse's heart. It seemed to me incomprehensible that the horse, after a brief, laborious effort, suddenly stood on trembling feet, then shook with revitalization. It put us all in a wonderful humor, and Ralph shook Marty's hand vigorously as I gave him a hug, and we all shared for that moment a very deep satisfaction.

But the appalling condition of the animal was inhumane. Lash marks showed all over his gaunt body. "Someone's really worked this guy over," Ralph said, wincing. "If I ever see that revolting man again, I'll have his head. Look at this pathetic mess!" The horse was hundreds of pounds underweight, and his teeth looked as though they hadn't been floated in years.

"Old boy's been starved to death. He can't chew. And look at the length of those teeth, Marty. He's older than I am." His hooves were so long that great, bowlike curves arched his legs. Inside his swollen sheath were open, painful canker sores covered in crusty grime. He had no tail, and his mane had been completely chewed off. There were gashes from teeth bites on his crest and flank. The only thing he appeared not to have was distemper.

We endeavored from that moment on to restore him to his original state, whatever that may have been. It was hard to tell; right now he looked like Ichabod Crane's horse, Gunpowder. The horse became a company project, and we labored over him slowly, not taxing him any more than necessary. First his teeth were floated, then his hooves trimmed, his sheath medicated and cleaned, and the sores on his body scrubbed with antiseptic. He stood there, that lovely old thing, enjoying every moment of our concern.

I walked him gradually to the barn. Sore and stiff, weak and shaking, yet he whinnied and snorted with long-forgotten sounds, then rolled around like a colt on the fresh-smelling, sawdust-filled floor. These were for me the special kind of heart-warming moments that made my life so wonderful.

I called him King, and a splendid communion began to take place between the two of us. Each morning I went to the stable to hand-feed him his vitamins in grains. As the weeks went by, I watched with great joy as his black coat began to glisten and shine. His mane and tail began to grow, and the curve in his legs straightened. A fine, proud crest began to form as weight slowly filled out his sixteen-hand frame.

I turned my spirited friend out into the corral attached to his stall. He rolled and jumped and kicked his legs far behind himself, delighting me for nearly an hour. Then, with a prance and a gait and toss of his head, he came to where I perched upon his fence and shoved a velvet muzzle into my hands. I rearranged a clump of new-grown bangs on his forehead and breathed in his intoxicating odor.

Suddenly I became aware of a familiar presence, one I had known from a distant past. Humbled and awed by the manifestation, I thought, "King, has time sent you searching for me, and now does the wandering end?" And I recalled a passage from a Walt Whitman poem: "Ages and ages returning at intervals, / Undestroyed wandering mortal."

Nearly two months went by before I rode the horse. Ralph pleaded with me to let him ride King first, in the event he bucked or tried to run away with me.

"Toni, you don't know anything about him!"

"I know enough." I cinched the saddle, took a deep breath, then climbed on that back all black and shining, my heart fluttering, feeling a "first-time-of-anything" excitement. Ralph held the reins. "Now you be careful, young lady; at least ride over in that soft area so if he should dump you, it will break your fall and not your neck." He led us to the sandy wash behind Swan Lake where "Nyjensky the terrible" was keeper of the pond–*his* pond, and if anyone so much as came within ten feet of it, he was sure to lose his kneecap from that rampant aquatic swan. We swerved to avoid his wrath.

And now, atop my marvelous King prancing in the sand, I said, "Okay, Ralph, let him have his head," and I gasped like a jockey at the starting gate. As I pulled back on the reins, King's massive head tucked in, and a majestic arch appeared in his noble crest. He was suddenly filled with pride, and I as suddenly was in the presence of an absolute monarch as gradually we put faith in one another and walked the old creek bed, haunted by the ghosts of the famous twenty-mule Borax team. On air we rode, soaring, soaring. We turned, King and I, superior beings, and loped

We were such dreadful snobs, the King and I.

back to where Ralph stood, showing his pride in us. At that time, as I sat astride the masculine black stallion, a feeling of femininity such as I have never known swept through me. It was a breathless and timeless moment.

As I was about to dismount, reluctantly, I eased back on the reins, and King reared straight up into the enchanted air. As though in slow motion, he rose on his hind legs, then gently placed me back on earth. I looked at a surprised Ralph. "Do that again—easy, easy does it," he said. Back again I pulled and once again, up we went.

"Why, someone's put training into this horse! That's enough for today.

We'll give him more time to collect himself, and then we'll see how much he knows. This is no ordinary horse." Of course I knew King was no ordinary horse. I'd known it all along.

Now every lovely day King and I went for short rides, and we began to build up his strength. His physical power became an energetic thing. The magnificence developing before my eyes was incredible, and everyone began to comment on his great beauty.

From the beginning we had grown possessive of one another, and other than Ralph, King would allow no one but me on his back. One by one he tossed all those who tried. It soon became known that King belonged to Toni and she to King, and no one disputed the fact. We were now spending more than a little time together, and I think I detected a pang of jealousy from Ralph. I was femininely glad.

As the months flew, we ran out of commands before King ran out of performance, and there was no lingering doubt that he was a very special animal. The mystery continued. The more we knew him, the more curious we became about his past.

The desert turns toward its few months of splendor each year in May and June when the wild flowers are in bloom and the poppies set the hills ablaze in Gaugainian brilliance. Yuccas' floral necks stretch toward the burning sun and thrive. While we dwelled within the spring bouquet, we were creative and industrious. The ranch shone with sparkle and polish. And sometimes, like the past state of the earth and encompassed by warm evenings, we rode naked and bare and fell in step with the decline of winter in the sand and sage. And the world was the color of a lion's eye then.

On one of those days of perfection, when all things are in harmony with one another, an old Hollywood wrangler came to the ranch and took a leisurely stroll back to the stables. In a few minutes he came running up to Ralph and me as we played with a young cross-eyed African lion that had just been given to us.

"What's wrong, Johnny, is something loose?" (That was always the first thought.)

"Nope! Nope! It ain't nothin' like that," he gasped. "It's that black stud back there. Where did you get him?"

"Do you know him, Johnny?" I anxiously asked.

"Yep, I sure do, there's no mistakin' that horse. But first tell me how you got him."

We told the story, taking shortcuts to hasten the end.

"Damn, if that ain't a shame! Well, my two young friends, do you know who that horse is?" He wrung his hands in anticipation. We were both bursting.

"Why, that beautiful old movie horse is the famous Fury!"

For years to come, Black Pegasus and I soared on dark wings, high above the minds of man, drank from the sparkling fountains of Pirene, and rose to the summit of Mt. Helicon, where we remained in a world of fantasy for centuries and centuries, established immortal. Having found one another, we were free. And so the legend grew.

6
The Animal Man

For thou shalt be in league with
the stones of the field:
and the beasts of the field shall
be at peace with thee.

–Job 5:23

1

Kenya sat wriggling in my lap; I rubbed his great ear while he contented himself with grunts of satisfaction and squirming detachment. I rather fancied myself as knowing personally every hyena who roamed our globe and felt passionately involved in bringing them out of the kind of persecution that made them out to be vermin and scavengers and hideous creatures. I was laboriously intent upon destroying the misconception described with such bias by Osa Johnson, who said hyenas were the one animal she shot for the fun of it, and by the likes of Kipling who referred to them as "those we hate."

I had aspiring notions of single-handedly showing to the world how sinister and lacking in truth these teachings were. So I invited everyone who was anyone to my little den of demonstrators. I pointed out the loveable nature of the animals with regard to man, and spoke of the dreadful examples of complete misconception that had been passed down from father to son for countless generations. And as always, whenever the cause

I fancied myself as knowing personally every hyena in captivity.

symbolizing the contradiction raised its challenging head, I raised my gun in retaliation. And, having taken careful aim at my opponent, to my horror I usually found I had in fact neglected to clean my own weapon, an oversight that sometimes resulted in regrettable backfire. It took some while, but eventually I learned to point the barrel in my direction, and soon its backfire helped me to become the undefeated marksman, muttering prophecies and moving ahead with eagerness to clear my own way.

Since no one felt about the hyenas as I did, we were seldom accompanied by human companions and so learned to rely upon one another.

Ralph had warned me time and time again about working with animals alone. But the argument had never quite penetrated, and since I had gone for years without even so much as a nick I was reduced to accept the fact that my condition was one of matriarchal power and that no animal in his right mind would dare bite me!–The word was out.

When the seedy Tinker Johnson drove into the ranch, pulling behind his straining old truck a dilapidated van holding a menagerie of pacing vitality, we greeted the traveling zoo with less than enthusiastic zeal, for the scene had a stifling effect on us all. Tinker had no ethical scruples governing the way he conducted his unorthodox trade. Animals were for him simply a way of making a living. He never regarded them as God's creation or fellow creatures, and certainly never did he look upon them empathetically. Tinker Johnson engaged in his life as an animal dealer with total detachment and no conscious conflicts.

I glanced into the dilapidated dwellings and had only to look at the stiff carcass to know the old bear was dead. Tinker was just emerging from his fermented surroundings as my wild look of astonishment had its usual stupifying effect on those round me.

"Missie, you don't look none too well. Why, you ain't got no color at all. How long you been ailing like this?"

"How long has that bear been dead?" I countered.

"Well, let's see now, I came down the mountain two days ago and he was having a kind of fit like, all stiffed up, and damn well mean, shakin' 'n foamin' like, made me sick just ta look at 'im. Musta died last night I rekin. I've been driving fer near twelve hours. Kinda lose track, ya know. Well, damn. There goes fifty bucks! Bugger just up and died on me, huh?" A growing impression swept through me that Tinker felt cheated by the bear's unexpected death. Since our opinions about animals were remarkably opposed, my expression communicated little. So I said the obvious.

"Let's go bury the poor thing, Tinker."

Had Ralph been around, a great medical scrutiny would have taken place. But he wasn't, and I, left to my own means, decided the only consolation for the departed was a sympathetic funeral. And so the martyred animal was laid to rest.

There was a sharp pain in my head when I returned to the truck and came face to face with a spotted Hyena, who, from his dimly lighted interior, peered at me with limpid, questioning eyes as if to say, "How could you, the one who knows, allow this to happen?" It's difficult to ascertain how many animals have said that to me, and always, unless I did something about it, I felt on the brink of doom. To save myself from myself I gave Tinker Johnson fifty dollars, but not until he had told me a heart-breaking story of an old widow whose last hope for her pet hyena was that he come to live at Africa USA. The boys shifted the dog–as I always called hyenas and my other caninelike friends–into a transfer cage and took him to the nursery, where they released him in the outside pen.

When the used-animal trader drove off into the bright red sunset, a sense of personal condemnation flooded through me as Ralph drove in because among all his lessons I had chosen to ignore were the loud words, "Don't ever buy from the likes of a Tinker Johnson; all it will result in is his support, and it will encourage him to trap more animals."

"But–but–that was a hyena in there; and it seemed some moral imperatives were operative–and–"

It was clear to me I was not getting through, for no one said a thing but me, and I was having great difficulty at that. Ralph smiled down on me from where he stood before the fire, hands in pockets, one eyebrow half raised. I felt like a criminal come to confess before God. The terror intensified as I searched the room for some verbal reinforcement. It was then the news of the dead bear was brought up. Until then Ralph just stood nodding disapprovingly as I pleaded my case.

"You *what!*"

"We buried a dead bear that belonged to Tinker Johnson," Ron announced in a fading voice. I began to shrink in my chair, knowing full well who was about to get it. Composed, yet with a funny color creeping over him, Ralph asked with terrifying dignity, "Where, may I ask, did you deposit the corpse?"

Having trouble swallowing, I answered, "He's laid to rest back by the storage pile. Why, what are you going to do?"

"Do? Not I. You–what are *you* going to do? You are going to dig it up, that's what you are going to do."

"I'm what! Don't you think that is a little macabre, I mean dead is dead you know, and besides that is a perfectly disgusting thing to say to me. I can't imagine why I should have to do that."

"Why? I'll tell you why–because you don't know what it died from."

"Well, it died. That's all I need to know, for heaven's sake. It just died."

With the mortician standing over us, we hauled up the corpse. The coroner, dressed in surgical rags and wearing long plastic gloves and horn rimmed glasses, performed the autopsy. I watched with complete disgust and gnawing reservation, then passed judgment as only a prejudiced jury could. "Sick–sick. That's really sick!"

But when the results came back from the lab, the devastating news was that the bear had died from rabies, and four of us would have to undergo the Pasteur treatment because we had come into direct contact with the animal.

–That's one, Tinker–.

2

Do you have any idea how much I loathe hugging and kissing people I remotely know? A lot! So then how come every time a new animal arrives I throw my arms around it and spout mundane sentimentality? Ralph told me once he was going to recall my lips. (You sly dog, you don't fool me. Oh no, *another* kisser!)

A box arrived from Africa by special delivery one early morning. There were air holes all around it, but they were so small I couldn't see in; so I set the box on the hand-made log table in front of the perennially shedding couch. I brushed my couch, Boris, once a day with a Hartz grooming device. Boris was usually covered with more hair than the animals who lived with us. Once Ralph sat down and asked with a note of detachment–as he examined the long collar I had fastened around Boris's arm–

"What's this?"

"Kennel Tags."

"Kennel Tags?"

"Yes, you know, I.D., rabies, dog license. The animal regulations chap was by and said if that 'thing' was to continue living with us he'd have to have his proper shots and be registered with the county."

"Toni," he said, his shaking head now resting in his hands.

"Yes?"

"Never mind."

–Where was I? Ah, yes, the box on the table. I couldn't stand it any longer. I had to open it. Ever so carefully I undid the straps holding the little crate together, and then slid back the top just an inch. I had learned early in my marriage always to be suspicious of cartons bearing moving parts. Ralph loves to tell "remember when" stories about friends of his who survived horrible ordeals.

There was the ever popular one about the man who lost his foot at Jungleland when he walked by a tiger cage and pushed back the meat that was protruding from the feed slot. Cheerfully Ralph told of a bizarre person who lost his nose while his chimpanzee was thinking up painful ways to take him apart. And of course there was always the charmer about the keeper who was beaten to death, then impaled upon the ivory of his favorite elephant. But the one I called to mind as I slid back the hatch was the vision of his friend, who, having received a package in the mail, couldn't wait to see what was in it and set immediately to opening the box, expecting a present from his mother for his birthday. He instead got a surprise from his rejected girlfriend. As he took the top off, a King Cobra popped his hooded head up, and the last Ralph saw of the man he still possessed the funny habit of swaying from side to side.

I peeped in the slit and was greeted by a little chirp as a fuzzy, bristled patch of baby cheetah "spooted" me. I threw open the hatch and grabbed the startled inhabitant with huggy huggies and kissy kissies. "Oh no, I can't stand it. You're so cute!"

"Ralph! . . . . Ralph! . . . . You won't believe it," I cried later as from the window of my station wagon I held high the sweet cub, then returned him to my face. We rubbed cheeks and I kissed his nose.

"Toni, what do you think you're doing?" he said with a note of disgust.

"I'm kissing his cute wittle nose."

"Where'd he come from?"

"A box."

"Well, where'd the box come from?"

"Africa."

"Who's it from?"

"I don't know; there wasn't any information on it. Here, don't you want to hold him?" and I extended a loving handful.

"Now, you know better than that. Take him to the clinic and have Marty check him over, right this minute. I'm surprised at you."

"He's surprised at us," I mimicked and began another series of huggy huggies-kissy kissies, then rolled up my window with some irritation.

"I wouldn't let you hold him if you begged me now. Come on, Squirt, let's go see Marty."

"Toni?"

"Yes, Marty?"

"Didn't you notice these round spots?"

That was the beginning of a new trend in my life: "There's a fungus among us." I broke out in ringworm the way some teenagers break out in zits. Seventy-five itching patches found their way to my body. For a while I was a walking contagious skin disease. "Eeew, Yukki! Look at Toni!" I had an uncontrollable urge to suddenly huggy-huggy kissy-kissy everyone I remotely knew. . . . I was released from my childhood phobia.

3

I knew as I walked into the pen with the new hyena that I shouldn't be there, regardless of the fact that I was prepared for the worst, wearing leather riding boots so high I couldn't bend my knees and a huge heavy army jacket—the kind one can survive arctic winds in—not to mention a football helmet and the thickest pair of engineers' gloves made. The experiment was to demonstrate to the world the possible validity of "laying on of hands." It was a beckoning challenge, impossible for me to resist. Then there was the overpowering appeal of proving my theory—that one could comfort and pacify another by touch. The idea had been proven with humans, but as yet no one had attempted it with wild animals. And here before me was the supreme test. Of course, I was by now convinced there wasn't a hyena in captivity I couldn't communicate with. So again, lulled by my own words and encouraged by the thought of all those incredulous faces watching me soothe the savage beast, I closed the gate behind me and took slow, well-paced steps toward the adult creature who appeared oblivious to my every move. With each forward advance I thought continually of my mental objective. My meditative state seemed one of detachment from my physical as I came even closer. The hyena now turned his head toward mine, and our gaze met. I felt I was doing quite well until a momentary suspicion engulfed me. But upon an in-depth observation of his penetrating stare, I knew contact had been made; and as I offered my gloved hand in friendship (so the dog could sniff and register my smell), it dawned on me that he couldn't possibly relate to my controlled energy field while it was bound and tied like a mummy. I removed the gloves and the jacket to show good intentions. As I did, I noticed his face begin to twist and contort, and a sparkling gleam of white teeth showed themselves in a faint smile. A rumble like that in a collapsing coal mine came from his massive throat. By now I knew how the game was played, but I had left my mind open too long, and my brains fell out. During that brief standoff the knowledge sent forth by the dog surged through me that I was on the verge of something far more than a social occasion. My eyes shot from side to side looking for an exit. His never moved from me. A tiny word escaped my lips . . . "Help."

I took one step backward, breaking the fierce concentration. It was my second mistake. My first was being in there at all. Averting my attention to the gate, I noticed the dog now beginning a series of mighty bounces up and down on a single spot. He was speaking foul words in a mongrel language. Fighting an impulse to run, I stood my ground, eyes bulging, temples palpitating, mouth hanging open. It seemed I had lost all control of the situation. I was not having fun.

From his throat came a suspicious noise of high-pitched defiance, and I knew then I had had it, for with unnecessary violence he came at me. It was a grotesque sight–a fierce upward thrust that caught my arm. He began gnawing on it as though it were an ear of corn. It felt a great internal discomfort as well as a colossal smarting sensation. Now, I cry at movies; I cry at parades, weddings and funerals; but I don't cry in pain except out loud. I began shrieking.

His mouth had me at every possible disadvantage, and I felt as though I were being ground to death. I tried biting him back, holding his nose, poking him in the eye (how quickly we forget), when from the corner of my eye a flash–like that of a sword–and from the pages of an Arthurian novel Lancelot appeared–one of my students–armed with a wooden cane. And much to my horror, he screamed out "That's it! Curtains!" and began beating *me* with the stick as though I were the one attacking the crazed dog. It seemed to add an element of poetic justice to the affair. My rescuer had seen fit to go berserk, and while trying to stuff his leg in the hyena's mouth, he let us both have it with the cane. I managed in my astonishment to let out such a blood-curdling cry that the planet stopped rotating long enough for me to push the startled man out the open gate. With a sickening thud my battered body fell on top of his–hurting him a lot, I hoped. My weight had a unique effect on the startled form beneath me, for as I hoisted myself up, he began shouting triumphantly, as if suffering from amnesia, "Hey, we certainly showed that bugger a thing or two!" I could only seem to repeat myself in a skip of surprise, "We? We?"

After locking the gate I staggered in the direction of the cheerless crowd now coming my way, led by my expressionless husband, who, by now (I was sure) was cursing the day that in a moment of male weakness he had seen fit to marry a lunatic. The worst part about that ordeal was never having a scratch nor a drop of blood on me to show for all that purifying pain.

A patient in a body cast could not have moved as painfully as I for those next few weeks. I had expanded in size until I looked like a balloon in the vague shape of a human. In that condition I reminded everyone for a long while of my past egomaniacal, self-indulgent state.

–That's two, Tinker–.

4

A log was burning in our old rock fireplace. Shadows danced about the warm cabin walls, and the flames lit Ralph's face like a sunset glowing upon the land before disappearing over the western horizon. Illuminated by the shining blaze, he sat propped against the dozing Zamba on the worn Navajo rug. Papers, spread all around him, continued the day's business long into the quiet night. I hugged blissfully to myself the comfortable scene and gazed around me, captivated by the soothing atmosphere of my home. It brought to my contented mind a revival of youthful aesthetics, of a happy childhood and a memory of irrecoverable innocence.

The sublime stillness was broken by the cry of the phone. Ralph said, "I'll get it, Honey." Standing slowly so that he would not disturb the snoring giant at his side, he tiptoed to the hall and lifted the receiver.

"Hello? Hello? Speak up, I can't quite hear you." After he had listened for a moment, he put his hand over the mouthpiece and whispered, "It's for you, Toni."

"Who is it?" I inquired. Ralph shrugged his shoulders and walked the long extension cord over to where I sat.

"Hello? Hello?" I was about to add, "No one seems to be there," when across the line came, "Aaaaaah . . . hha . . . ahhha . . . hhhaaah . . ." It was a "breather," hyperventilating into my ear. I slammed down the phone with astonished distaste and, lifting my eyes toward my poker-faced husband, I said calmly, "For me? That call was for *me*?" then added with offended inspiration, "I won't forget you for this."

"I've got a surprise for you," he said, inviting a change of subject.

"Yeah? I just had one of your surprises, and I don't think I'm ready for another one just yet."

"This is something I think you will like."

"Oh, all right, what is it?" I winced.

"Tomorrow you're going to be in charge of a studio job."

"What? *What*? You don't mean it." I began to swell, lifted by excitement. Then, seized by uncertainty, I asked, "Do you think I'm ready?"

"Do *you* think you're ready?" he returned. I took a long, deep, prayer-filled breath.

"Yes, yes, I can handle it. I'm sure I can handle it."

After going over with me the scene in the script for an episode in *The Virginian,* he said, "Let's get some sleep. We've both got early calls tomorrow."

"You mean you're going with me?"

"Nope, you're on your own. Don't worry. Ross and Jamie will be there with you." With that he gave my icy face a reassuring kiss, and I began to melt, for when Ralph kisses me, I stay kissed for days.

Four a.m. came way too soon. The discomfort of rising on a black, cold morning had never been a pleasant way for me to start my day. Wanting to eject the contents of my stomach, I cleaned myself up, passed on breakfast, and blindly made my way down to the ranch. "I must be the only person up in the whole world," I thought. But when I drove up to the cat circle, Ross and Jamie were already loading Tom (the mountain lion) and another cougar into the back of the camper.

"Hi guys, what can I do to help?"

"You're a little late, Toni." Jamie's voice was cold. "Everything is loaded and ready to go."

"What do you mean I'm late? It's only 4:30. We don't have to be at Universal until 6:00." He threw me a frigid look.

"Come on, let's go." He held open the door to the pickup so that I could sit in the middle.

"Hang on just a minute, Ron. I don't know what your problem is, but I'm not going anywhere until I've checked to see if we have all the tack we need."

"We've got it; you take my word for it. Or isn't my word good enough?"

"Hey, come on, Jamie, I know it's early, but that's no reason to be so sharp. I'm ramrodding this job today."

"Yeah, I know."

"Well, like it or not, that's the way it is." And with that wonderful tension-builder, I sorted through the trainer's tack box to ease my mind, checking off mentally the equipment we would need: three catch ropes, three leashes, double clips, clevises, rings, canes, one bucket of cut-up meat, CO_2, axles. And to that I added my own little tool box filled with pliers, hammers, monofilament spray cans, an assortment of unfeminine overalls, and a prayer. Both mountain lions were sound asleep in the hay-bedded transfer cages. I shut the back doors and came around to the front of the truck, climbed in past a resentful look, and sat for the entire uncomfortable ride between two aggrieved men who were (with growing reservation) working that day for the boss's wife.

It was hard for me to establish myself in the business because, among my other handicaps, I was a woman. In the world of professional animal men that was a definite drawback. *Female* usually meant *weak* and *fragile,* and that meant the men had to be protective of me when I was working, more than they would another man. As far as I was concerned, that brought out one of their better qualities; I don't believe anything is quite as masculine as a man concerned about a woman's safety. But I could see how it would add one more anxiety to the already nerve-wracking job. So for nearly a year I had audited everything that went on occupationally. I had entered a man's world, and no one was going to force me on them. I would have to prove myself in a career where I wasn't particularly wanted, not only by those in the profession but by the studios as well. They had their confidence already established in enterprising young men, and well they should have. And here I was, the wife of the new "prophet of the beasts." And I felt completely isolated by intimidation and neglected learning.

I must have driven Ralph crazy that first year as I followed him around like a lost soul, who, although grown up, had never gone to this kind of school. I was so interested in this fascinating, futuristic life that I'm sure my curiosity and I drove him to distraction, casting a shadow atop his own as I did then.

When you are suddenly confronted with the moment of proving yourself, it can be a time of terrifying awareness. You wish you had never made such brazen statements as, "I'll be the brains behind the brawn," or "Oh yeah? Well, it doesn't look so hard to me."

The throbbing of my heart found me on my first studio job, where I was expected to live up to all the obnoxious remarks I had made at the ranch. I knew I was on the line, and the one thought that rose above all others was, "I can't embarrass Ralph. I'll never be able to face him again if things don't go well today." So, lifted by the wave of something worse awaiting me at home, I recovered long enough to face Ross, who asked in a cynical voice, "Well, Boss, how are we going to get this shot?" And his eyebrow raised as he lowered his head quizzically and gave me a villainous smirk. What was I doing in this business? I didn't even like animal movies.

I felt my face flush and my nails dig into my sweaty palms. "You don't think I can handle this, do you Ross? I'll even wager you've taken bets that I'll fail today." He shrugged his shoulders and smiled like a naughty little boy caught with his hand in the cookie jar.

I felt my eyes fill as a sensation of defeat came over me. "Oh, Ross, how could you? I just think that's awful. I don't want to take your glory. I just want to do a good job." And I burst into tears.

Ross, moved by my misery, put a reassuring arm around my shaking shoulders and soothed me the best he could, considering he was the enemy. After a few more of my sobs, he announced with a certain cheerfulness, "You do realize you're in my arms, don't you?"

I exploded away from him, snarling, "Don't you ever think of anything else?" And I kicked him in the shins.

"Not if I can help it," he replied with a pained grin, hopping on one leg and holding a swelling ankle. I began to laugh as he began to laugh, and a sudden transformation of character occurred. "Oh, look Ross, I'm really sorry I kicked you. I'm not trying to hide anything. You know and I know that I'm not a trainer. But I've got to start somewhere, and I can't do it without your help. So please, won't you humor me if nothing else? Like it or not, I'm going to be around for a while, you know. I only want one thing out of life, and that is to make Ralph proud of me. Is that so awful?"

"Well, of course not. Come on, Kid, don't take everything so seriously." And then the call came across the megaphone, "Where's the animal man?" and Ross cried out, "She's over here."

Odd, isn't it, how one can be so self-assured one moment and so completely lacking in confidence the next? Whistling a few bars of a tune I made up, I walked toward the assistant director. I did all I could to stop my teeth from chattering, so that if nothing else I could at least look as though I knew what I was doing. The assistant director looked like a nice enough person. But he shattered all my expectations when he said, "You're the animal man? Uh ... Uh ... woman. Well, that's a good one." He laughed. "You look like a little kid." And again he laughed. I failed to see the humor and said rather stoically, "If you'll be so kind as to show me where we are to film this scene." Then I added a few well-seasoned technical terms to our conversation until he began to shake his head approvingly, and we set to the business at hand.

Working as much on the will of others as on my own, I carried on as I was sure the wife of the great Animal Behaviorist should. But having no comparison to draw from, I made a lot of it up. And Jamie watched my act with less than approval, then asked in a lecturing manner, "Is it your time of the month? We're working with big cats, you know."

"Smooth, Jamie, smooth. You really know how to make a girl feel right at home. No! It's not, and if you had any class, you would have checked the chart at the office. Now, without any further discussion, let's do our job, shall we?"

"Yes, ma'am." He saluted me, standing at rigid attention. Scarcely had I reprimanded the man than I felt suddenly inflated by my own authoritative words. I was seized by an undeniable faith in myself. I was Mrs. Ralph Helfer, for heaven's sake; that in itself was an achievement of the utmost wonder, and as Mrs. Ralph Helfer everyone (outside of my own little group) would expect that I knew exactly what I was doing. Why, the name alone gave me unquestionable authority. I had been trained by the leading expert in this field for over one year, and he had drilled into me technical knowledge that had taken him a lifetime to develop. I was his star pupil and most enterprising subject. Was there any doubt? Of course there wasn't. There was nothing I could not do! And today I would show to the world how fortunate the animal professionals were to have me among them. I had cheered myself up to the point of near brilliance, and, feeling free from weight, I began to set up the shot with fanatical haste. This was certainly one business where fat people could not exist; between the anxiety and the work load there was never an opportunity to indulge in the luxury of excess pounds. My body was as hard as a rock and sometimes so was my brain.

Resolutely I went about my task of unloading the truck and issuing orders to my chauvinistic crew, indulging myself in fanciful daydreams: about the excellent job I was about to be responsible for and about accepting congratulations from the director, blushing with proper modesty as the group of studio people applauded my deed with enthusiastic rounds of admiring applause. And when the producer made his way through the crowd to award my efforts with a bountiful bonus, I would be so overcome that few words would escape me. And of course from that day on Universal Studios would never dream of filming an animal sequence without the legendary T. R. Helfer carrying on the family tradition by directing such activity.

During my reverie I must have been wearing a somewhat strange look, for Jamie, who was helping me unload the mountain lion, broke the spell with his usual thoughtless taste. "What's wrong with you? Ya got indigestion?" I ignored that bit of sentimentality and kept repeating under my breath, "Focus, focus. We must have focus."

Ross was doubling for the actor, and Tom was playing the part of a wild mountain lion, who, seeing man intruding into his region, lay in tense anticipation for the opportune moment when he could jump him from his ten-foot vantage point.

The plan was for me to release Tom from the overhanging ledge. Ross, already in wardrobe, would give the celebrated wrestling cat his cue to jump on him, and Jamie would pick up Tom with his leash down below when the cameras had stopped. Now that sounds like a simple enough scheme, uncomplicated and orderly.

Tom was an extraordinary cougar, friendly as any living thing I have ever known. He had a deep regard for people. The more he was in our company, the more he seemed to flourish. One had only to scratch him under the chin to send him into rapturous purring, and his amber eyes, framed by tawny fur, would close in rapture as he enjoyed man's touch.

From the time he was a small cub he had loved to wrestle, and, having reinforced this natural behavior, his time was for the most part spent in tossing the men at the ranch about. And as these were still the days of the western cowboy epics, Tom was forever fighting to maintain the wilderness as animal territory.

Tom and I had climbed up the back of the small cliff, Tom tripping me all the way; his favorite walking position was between my legs. We squatted down just out of camera range, and he rubbed against my shoulder and licked my arm, purring in a low, vibrant sound, completely contented. I uttered small, high-pitched cougar imitations that could have been mistaken for a bird.

I called down, "Whenever you're ready, Ross," and the director shouted, "Cameras rolling, speed–action!" I turned Tom loose as Ralph began calling him, and I slid on my belly backward, clinging to the hillside to stay out of sight. Tom stood, walked to the ledge, and peered down. I heard both Jamie and Ross coaxing him. "That's a good boy, come on, Tom. Get me. Come on, at-a-boy," and I knew Ross was running around below in circles and tempting Tom by turning his back on him and crawling about on the ground. But Tom kept looking over his shoulder at me. And now his head was sinking down, and he crouched low, his front paws digging for traction in the earth. I was not enthusiastic about what was going to happen next, but there was nothing I could do about it. "No, Tom!" I called in an authoritative voice, shaking my finger to head off the inevitable. With devilish, laughing eyes, he came flying through the air at me. "R-O-S-S!" My cry was muffled by a struggle to master some ground. Each time I tried to stand, my knees buckled under my gymnastic opponent. Two competitors more mismatched have rarely been seen. Finally I screamed, "That's *not funny*!" And Tom stopped long enough for me to change the subject by scratching him under the chin. As his mood changed to ecstacy, he lifted his loving head and closed tight those playful eyes, and I didn't dare stop rubbing; I knew the match would begin again. And that was how they found me, Jamie and Ross, appearing as though I was having a wonderful time spoiling their shot.

"–Ah–I suppose you wonder what we're doing? Well, I was–and Tom somehow reversed–Oh, Ross, this is ridiculous! You'll probably never believe it but Tom (who sat angelically by my side, purring as I stroked his chin) jumped *me* instead of you!"

"Well that was dumb, I mean really dumb!" Jamie exploded.

"You must have done *something* to change his train of thought!"

Having put a leash about the great cat, I stood, stretching to my full size, towering half a head over the Napoleonic man who saw fit to deliver me a vicious insult at every opportunity. "Listen, you nasty, biased, pompous little creature, not only are you small of stature, but you are small of mind. You have issued your last mouthful of contemptuous remarks at me. I am not your victim. At least not during our professional time together!" (If there is to be any more, I thought to myself) "Here!" And I handed him Tom's leash. "You release him this time, Smart Boy. I'm going down below." And with that Ross tiptoed out of sight, and I hissed a last exasperated gasp at Jamie. My whole body trembled at the thought of what I was being called under his breath. I walked down the hill slowly, allowing myself time to recover. Ross, in fear of being abused by my out-of-control tongue, looked upon me as if I were an active volcano and kept his distance.

We took our positions as the director, bewildered, asked if everything was okay. "Couldn't be better," I replied as Pollyanna might. "All ready when you are." The A.D. called "*Quiet on the set*–Don't anyone move!" And we were off for the second take.

"Tom! Hey Tom–come on Tom, at-a-boy," Ross yelled as Tom peeked his lovely head over the edge of the rock to see what kind of strange human madness was taking place just below. Ross was running around like a mechanical wind-up toy in need of oil, occasionally falling on the ground and scooting around like an uncoordinated reptile out of its element, trying to arouse Tom's interest. Just as the playful mountain lion looked as though he were about to jump on Ross, he took a side glance around him and leaped backwards out of sight. I heard a few inaudible words from up top, and the director yelled, "*Cut*!" Ross and I raced up the short hill to see Tom and Jamie rolling about in the very manner that we were trying to shoot. I had difficulty speaking; I was trying not to laugh.

"Will someone please get this wrestling maniac off me?" Jamie cried in frustration. I wanted to say, "Why, I can't possibly imagine what you could have done to make him do such a thing." Instead I gave a "Bad Seed" smile at Jamie while Ross placed a leash on Tom.

"All right, all right. So I'm sorry," Jamie spurted; "I was wrong."

I was so startled by his apology that I replied by saying, "It's all right; it could happen to anyone."

The A.D. and I devised a new plan of action. We fenced in the hill, camouflaging the chicken wire with brush, left Tom alone in that spot, and Jamie and Ross and I went down below. "Are we going to get *this* shot?" asked the A.D., glaring at me.

"This is it," I said with bells in my voice, trying not to show that I was praying to every god that had ever been mentioned, bargaining with fate even to the point of suffering eternal maltreatment from Jamie's verbal insults. Then crossing everything there was to cross (even my eyes), I said, "We're ready."

"All right, here we go again. Everyone quiet on the set. Don't move," the A.D. ordered with less enthusiasm in his voice. "Rolling cameras! Speed! Action!"

"Tom, come on Tom–come on–at-a-boy Tom." Tom peeked from above the horizon and stood for a timeless moment, gazing around at his isolation, then looked down once again at Ross performing his peculiar choreography. Slowly Tom's ears went back, his shoulders unlocked, and his head went down. As he looked up, lo and behold, he jumped, and a superb wrestling match took place. Ross popped a blood capsule in his mouth as a shot rang out, scaring the cat from him as our cowboy hero entered and Tom ran to Jamie's side. And Jamie, as proud as could be, patted and reassured Tom in a whisper that he had done a fine job.

"*Cut! Fantastic*! That was fantastic!" and Ross and Tom received a round of applause and a hearty handshake from the director. And I got a wink from Jamie, which meant more to me than ever being head trainer again.

Having wrapped it up and feeling an especially good sensation that I wanted to share, I ran to a nearby phone and called home. Laurel answered from the switchboard and said Ralph's set call had been postponed until later in the day. She put me through to the house. Ralph answered. "Hello? Hello?" and I started breathing heavily into the receiver. "Ahhhhh . . . aaaaaahhhhhh . . . ahhhhhhaa . . . hhhhaaa . . ." Ralph said, "Sorry, my wife's not at home." And I heard a click as he hung up.

5

Not too long after that first catapulting experience, I found myself with a little triumph. I was actually beginning to like reptiles, one in particular—Asteroth, a 14-foot, 75-pound Indian Rock Python. She was a "charming," sensitive snake who regarded me rather as a friendly tree. It was wonderful being relieved of a neurosis inflicted on me by a neurotic society. Oh, I still retain the memory of such a fear, but the fear itself was cured by months of wise counsel coupled with patient rehabilitation and educational therapy. When Ralph finished my schooling, I became one of those converts in search of people's unhealthy phobias, eager to pass on the "word" as I had heard it. Is there anything as provoking as a newborn convert?

When Asteroth was hired to work at MGM, I eagerly volunteered to do my first reptile job, and Ralph, proud as he could be at his pupil's graduation day, briefed me on how to get the shot. Because he was totally involved shooting the new TV series *The Greatest Show on Earth,* I would have to handle this call on my own. I had much to prove, so when I arrived at the lot that wind-whipped bitterly cold January morning, I wore Asteroth like a living boa, wrapped around me. In this way, my body heat would keep her warm. Over us I buttoned a heavy, long coat. I looked to the outside world like a very fat lady with a little pin head. We shuffled sideways through the sound-stage door. I felt a little breathless. Having just walked from the outside parking lot carrying seventy-five extra pounds left me in need of a place to sit before checking in with the assistant director. As I was about to pull up the first apple box in sight to rest my weighted bones, a thin, greyish, studious-looking young man wearing thick glasses, smoking a pipe, and carrying a leather-bound script, walked nervously up to me. His eyes, refusing to meet mine, stared intently at the floor. As he introduced himself, I stood up.

"How do you do? I'm Wilfred Crats. You are Toni Helfer, aren't you?"

"Yes, I am."

"Er, ah, er, ah," the assistant director sputtered, as he looked anxiously about me. "Where is it?"

"It? What *it?*"

"The snake."

"Oh, you mean Asteroth. She's right here." And I patted my sides as pregnant women do their stomachs.

"*What*!" Wilfred shrieked, as he jumped ten feet backwards. "You mean he's actually on you? How revolting!"

"How lovely", I thought to myself, "another of society's victims who desperately needs my help". I took a step toward Wilfred and was about to enlighten him when he took ten more steps backward, glaring wildly at me like someone possessed. He screamed, "Stay where you are. Don't you dare come near me! Don't you dare! I swear I'll do something awful!" His hand was twittering violently about, and he wore a twisted expression. Asteroth and I became immediately immobile as the poor man continued to tremble.

"But Wilfred, you don't understand–"

"Understand! You should have told me the thing was in here!" At that particular moment Asteroth peeked his innocent head over the top of my coat, just below my chin, and jutted his tongue out at Wilfred, who, seeing the snake, went completely berserk and fled at top speed, screaming all the way out the stage door. I stood bewildered, thinking how sick that poor man was. With a heavy sigh I pushed Asteroth's head back under my coat, just as a round of uncontrollable laughter broke out behind me. I turned to see a man wiping his eyes and blowing his nose. Propped against the wall, he began to laugh again. When he saw I was watching him with some question, he tried to compose himself. Through another series of "Ho ho, ha ha, hee hees," he managed to distinguish a sentence.

"Oh, I say, how absolutely splendid. That was absolutely the most splendid moment I've had in years. My dear girl, you have got to tell me, ho ho, ha ha, what you said to that fellow that sent him away like that? Frightfully funny, an absolute smash! Ho ho, ha ha, hee hee. Oh, dear. I must get control of myself. Please, do allow me to introduce myself. I am Dexter Poinbrook." Dexter Poinbrook? Wilfred Crats and Dexter Poinbrook in one day?

Dexter Poinbrook was, as you may imagine, English. And when he spoke, it was as one having had judicious training. He was an actor, a tall, handsome, distinguished, healthy man. I squared my shoulders and entered into conversation with him, not yet mentioning the presence of my clinging vine for fear of alarming him. I fancied bringing Dexter home for Stevie; he seemed to me a perfect type for a brother-in-law. There's something about an educated Englishman that constantly reminds one of how ill-bred most American males are, especially after one had just run screaming from my presence.

"You look as though you're freezing. Do allow me to get you a cup of tea; then you must tell me what that was all about. I simply won't be able to carry on until I know."

Still alarmed at Wilfred's emotional outburst, I couldn't help wonder what Dexter would think if he knew that beneath my coat was more than just me. He did not appear to be a man seized by fears, and I would not intentionally frighten him. If there was one thing above all others Ralph had instilled in me, it was to respect other peoples' fears and never use an animal to promote them.

"Tea time," came Dexter's cheerful cry. "Sit here and tell me what happened. Here you are, my dear, just the stuff for a brisk morning," and he rubbed his hands together after setting my cup down. Hardly had we spoken when across the massive room came the familiar cry, "Where's the animal man?"

I stood and waved across at the second A.D.

Ralph never told me I would be known by the company I kept.

"Okay, Helfer, you're on." Without thinking, I unbuttoned my bulging coat and mumbled to Asteroth that the scene was an easy one. We should be able to get it in one take and be on our way. Having momentarily forgotten about the aristocratic Dexter Poinbrook, I turned and met silence and a look of bug-eyed distaste. The man was stunned.

"What a revolting thing to do! I tell you, it's shocking! And bloody awful! Who are you?" he asked, as if I were a leper.

"Oh Dexter, I'm so sorry. I wasn't thinking. I'm an–" But so much for Dexter Poinbrook. He and his undiminished vitality had vanished from my sight before I could finish the sentence. Once again Asteroth and I stood in utter isolation, bereft of the friendship of others.

We walked toward the set as the crew now scattered in all directions, calling out indelicate words, climbing ladders, clinging to walls, holding one another. I felt almost like Moses walking the ocean floor as the Red Sea parted. How could I ever convert anyone if they wouldn't so much as come near me? And Ralph never told me I would be known by the company I kept.

6

Victim after kidnapped victim was sent to us from every corner of the earth, and for years we took in hundreds of orphaned wilderness refugees at our own expense. For a while it seemed to be in fashion for thoughtless people to dump their responsibilities over our fence. African lions arrived in damp crates with notes attached: "This is Flopsy. He only eats vegetables, and he sleeps with his Teddy bear. Animal Regulation won't let us keep him." Mountain lions were left tied to the gate, and once an elephant was delivered with no return address. Poor, frightened things, tossed about like so much excess baggage. Some of the worst offenders were actors and actresses who, having made films in Africa or India, came back with a tiger or a cheetah or lion cubs (it was the chic thing to do), made the circuit of TV talk shows, got good press coverage, and, when they had exploited the animals to the fullest to feed their egos and the novelty had worn off, they came to us. Then there were those less fortunate people who had no way of attracting attention to themselves, so they bought exotic animals to make people take some notice of them.

The great majority of "animal lovers" (and I use that term loosely) were people who suffered from lack of love themselves. They showered their animals with unrestrained affection, and that *always* resulted in spoiled monsters. These were the hardest hit casualties when they were given away.

It never ceased to amaze me how startled the people were when the animals began to grow. "Why, I had no idea it would get so big!" "It ate my dog!" "My neighbors complained about the noise." "It attacked my child." One after another the stories came in, and right behind them came the animals. There should have been laws against such things, but there were none then.

There came a point when we could no longer afford to accept "donated" animals. We also had to stop visitation rights to many of the more neurotic donors. They didn't want the responsibility, but they certainly wanted the control.

"Little Floppsy Woppsy doesn't look like he's very happsy wappsy without his Teddy Weddy bear." Sometimes I couldn't stand it any longer, and once I made the mistake of retorting, "Madam, why don't you go home and get pregnant!" The remark created a dedicated enemy.

After the paranoid "people haters" came the crippled minds, who, not accepted by society, took vengeance on their animals. Scalla came to us through one of those donors. No note was pinned on her narrow, decaying box, no excuse for the sadism that had been practiced on the box's tortured inmate. The only writing was on the crate: "Eats dog food." When the casket was opened, a tiny dwarf of a lioness fell from it. The look in her wet eyes was of excruciating pain, and when she raised her lip to snarl at us, we could see the exposed nerves on the tips of her filed fangs. Her tiny feet were so sore she couldn't stand. The pathetic creature lay on her side, hissing. The horror this little thing must have undergone put us all in a humble state.

Some time after corrective surgery, the pain left Scalla's eyes, but the fear and mistrust remained. Try as we might, we couldn't seem to break through to her. The brave little warrior never gave an inch. She was a confirmed people hater, with every right to be one.

We became concerned about her fate. She couldn't survive on her own in the one-acre pen that housed the other lions. She was deformed, weak, an imperfection, and they wouldn't have allowed her to live. She stayed in the nursery with me.

One month after her arrival, our devoted wolf couple Rhona and Hondo delivered their April litter. Rhona was getting old; she was sixteen now, and her interest in puppies wasn't what it once had been. She was content to be fully absorbed with Hondo, the great love of her aging life. These two were an inspiration to be around, and they taught me about compatibility. Through the years it had been a rare and fulfilling experience for me to be accepted by their children as one of the pack.

Rhona rejected this litter of four pups, and they were placed on substitute feedings in the nursery. At night I took the puppies home, but during the day they occupied the stall just next to Scalla.

At feeding time in this virgin country, as I gave the tiny clusters of howling fur their formula, there came odd whimperings, soft, intermittent sounds, from Scalla's room. I had been sitting on the floor holding one pup at a time as they nursed from their bottles, and as I turned my head, to my surprise I saw the amber eye of Scalla, peeking at us from the crack in the redwood divider. She made such continuing little overtones that my curiosity overcame me. One of the pups trailing me, I walked around to her front door and opened the top half. The little lame lioness was unable to jump very high; so I knew the pup and I would be quite safe from this vantage point.

Scalla, who had never ventured near before but had always cowered in the corner whenever man approached, came to where we stood, and, putting her two front paws on the top of the dutch door, squeaked and squawked, rubbed and sniffed, and licked the hand holding the pup. I held my breath. Here were no barking, hissing, and glazed wild eyes. She was passive and maternally involved. I opened the bottom door and let her out onto the runway, and she advanced toward me like a long lost friend. As I held the puppy high, she actually let me pet her! What a thrill it was to be touching this compliant animal that, until one moment ago, had been completely wild. After several days of exhilarating observation, Scalla moved in and took over mothering those puppies. I was an avid witness, fascinated and awed. Ralph had told me of such happenings, but this was my first experience–a revelation.

My lovely little Scalla–all the fear and hate left those beautiful amber eyes. One would never have known she was the same animal. Outwardly she didn't change much, although her fur became shiny and thick, and she put on a good deal of weight. But inwardly she was transformed, thanks to Kasan, Juneau, Thor, and Kemo–who showed me the way into the heart of my faithful, dwarfed friend.

Scalla raised that pack as though they were her own. I wonder what Rhona thought when she saw the lioness stroll by with one of her rejected pups relaxed and hanging from Scalla's toothy mouth. Scalla kept the pack meticulously clean, often lifting them into the air with great strokes from her long, rasping, sandpaper tongue. She reprimanded them when they were bad, cuffing them with her cushioned pads, sending them sailing head over heels through the downy soil. And she exercised her

As the wolves grew, Scalla remained small.

lungs at feeding time when the pups threw back their heads, closed their eyes, and howled in the ritualistic rite of their kind. At night I delighted in watching them as they slept in bliss, snuggled next to their warm, protective adopted mother. Scalla of the wolves was a patient, gracious lady, tolerating the mischievous nature of four growing canines, who used her back for a slide, her ears for teething rings, and her tail for a tug of war. Scalla's gentle heart had found a home. As the dogs grew, Scalla remained small. It wasn't long before they towered above her. But size had no bearing on their relationship; she remained the leader of the strange procession.

Two or three times a week we took a leisurely stroll through scorched grass and across mohave dust heaps to a green oak tree overlooking the

A small hill under our favorite tree.

ranch. As I sat, Scalla would settle next to me and put her golden head in my lap. Here we would relax and watch as the wolves terrorized the native horned toads and lizards. Wolves are so full of themselves when they're young. They continually trip over their great, gangling legs, and their piddly little tails are released when they submit to the "King for the Day." They are born comedians and carry a great sense of humor throughout their lives.

But because of their monogamous nature, I always found it important to have them in close contact with as many people as possible, both male and female, for when they become adults, they have a tendency to become extremely possessive and protective of an individual who has been too close to them, even though they may live with a pack. This was true with several of my adult dogs, and I had to be careful with them when we went for walks. If anyone came near me, up went their lips and ruff, and down went their heads, as their eyes became fixed and they growled, ready to defend me to the death. The world suddenly became tense and still. My big male Sancho grew to 150 pounds, and Little Red Riding Hood often had her hands full. To quote Mohamet, "Paradise is sometimes under the shadow of swords."

When Scalla died, the wolves and I felt our loss greatly. But Scalla's three years with them had been happy, and she died peacefully with her family around her. We buried her on the small hill under our favorite tree, and that spot became a meaningful piece of earth. For a long while after that, the pack and I would go and play with the horned toads up there, and the wolves would always sniff the earth she lay beneath. And Kasan, never as rambunctious as the rest, would sometimes lie next to her grave.

Rest in peace, my kind, warm-hearted little Scalla of the wolves.

7
Tana

The wolf also shall dwell with the lion, and the leopard shall lie down with the kid; and the calf and the young lion and the fatling together; and a little child shall lead them.

—Isaiah 11:6

Old Bessie, Ralph's truck, born in the '30s, was a grand old pioneer dame. She was eagerly engaged in work from dawn until dusk, and in all the years I knew her, her undying service to the company never faded. She was laden with responsibility; nonetheless hard work was her motivation. One had the impression that the indomitable and energetic Bess, with due dignity, fully comprehended how necessary she was to the industrious operation. She was indeed of crucial importance.

In her day there wasn't a worker to compare with her. She ran the errands, maintained the roads, fed the animals, delivered the hay, hauled the waste—and never once complained of overwork. Though she had developed a slight case of the shakes, and her cough had gradually grown worse, her heart—while skipping a beat now and then—was always in it. And the spitter-spatter of her faithful engine repeated over and over, "I think I can, I think I can," as on many a fog-filled morning she gave her all as the motorized Nanny of Africa USA. "I thought I could, I thought

The lion and the lamb. (Photo by Richard Hewett.)

I could," came the sound from her weak lungs as her strong new rubber wheels (a reward for service above and beyond the call of duty) carried her off down the sandstone road and into the 6:00 a.m. chores.

Brave Bess died a heroine's death in the flood of '69, saving lives at the loss of her own. Things were never quite the same without her, and the ranch never operated as smoothly. Blessed were we to have had Bess.

When I became pregnant, in my eighth month I had only a pot belly. If there was such a thing as *feeling* pregnant, it never happened to me. I swam nearly every day, rode my horse religiously, and never failed my latrine duty at the compound. I was never sick, and the only reminder that I was carrying a life was an occasional backache that was easily relieved in a warm tub. I paid as little attention to my condition as it did to me.

On the evening of June 2 I lay like a bellied-up frog, partially submerged in a hot tub (whence Jerome the alligator had just been evicted), when the phone shrieked. I ignored it, preferring to daydream about the big event.

The phone insisted upon being answered. I stood, dripping. In my gruff voice I answered. "Hello!" (Uh-oh–it was Estelle!) "Sorry, I didn't mean to be so quick."

Ralph's mother was a warm, caring woman whose whole world revolved around her children. They were the pride of her self-sacrificing life.

"No, Estelle, you didn't disturb me at all. I was just relaxing in a hot bath. My back's been giving me a bit of a problem today.

"How long has your back hurt, Dear?" she asked.

"Since this morning. Why?"

"Toni, I want you to call your doctor and then have Ralph get you to the hospital."

"Don't be silly. It's just a little backache. I've got two weeks to go yet."

"Toni, listen to me. You're in labor, Dear, now do as I say."

My first thought was that my legs needed shaving and my next that Ralph wasn't there, having taken the station wagon and gone to visit his sister.

"Ralph's with Sally!" I exclaimed. "And there's no one at the ranch. They all went to a studio premiere and took all the cars. I'm alone! Fifty miles from nowhere without a car," I quavered with mounting alarm.

"Now, Dear, don't panic. I'll reach Ralph at the studio. You call your doctor and then call me right back."

She had the situation under control. I dialed the doctor. Now I was mad. He hadn't told me labor could start in your back. I knew I should have had a woman obstetrician. What do men know about having babies, anyway?

Yup, the nurse confirmed I was in labor and the doctor would meet me at the hospital. Click, she hung up.

I called Estelle back. "Now what?"

"I've reached Ralph, Dear; he's on his way. But he's an hour from you. Think, Toni. There must be someone at the compound. He'd never leave it alone."

Ah-ha, she was right. Sam was there! "I'll call you back."

Sam was Ralph's father.

He wouldn't hear the office phone since he'd be making his rounds as night watchman on foot. Even if I reached him, he had no car. I had visions of delivering my own baby. If I had to take the *horse,* I was heading for the maternity ward.

I walked down the stony dirt road to the highway's edge and screamed across to the ranch on the other side. "Sam! Sam! Sam!" It was pitch black out. My voice broke through the silent night: "SAAAM!" Well, if he couldn't hear that, he must be deaf. I looked quickly around, thinking I saw shadows jump behind the trees.

"Ouch. Oh! If she hadn't told me I was in labor I wouldn't be in labor." I was perfectly willing to go through life as I was. Actually I'd become quite accustomed to my expanding condition. The thought of my stomach bursting began to terrify me. "Sam! Sam! Oh no, my yelling has started everything roaring." This was useless. "Well, Prince, you probably know more about delivering babies than I do. I guess we'd better go look it up in Mercks' Veterinarian Manual and boil some water."

Prince stared at me rather vacantly. He knew I was in trouble. He wouldn't desert me in my hours of need–even though everyone else had. He headed slowly back up the hill.

"Birth! Reproduction! What an ego trip, Prince, stamping ourselves out in duplicate, assuming more of *me* is needed. Whose idea was this?"

Prince barked as I opened the front door. "You're right. Let's sit down and think this over, calmly and intelligently." The only intelligent thing tc do was scream. "SAAAM!"

"What?" came a voice from behind me. I nearly had the baby right there, he gave me such a fright.

"Oh, Sam, thank heaven you're here. I'm in labor. We have to get to the hospital! Sam? Sam?" He looked catatonic. You had to understand Sam. Through the years he had shown up now and again with a new wife. But he always professed love for Estelle, who begrudgingly took him back between marriages. When Ralph began to make a name for himself, he suddenly became the apple of his father's eye. Sam clung to Ralph's success like glue, and became dependent on him overnight. Ralph bought Sam a trailer and put him on the Wilson's ranch, where he could fulfill his lifelong ambition of raising frogs.

Between Sam and Ross, no one knew for sure who owned the ranch. At least Sam's embossed business cards had style, while Ross's were black print on cheap white paper. The identity crisis is a pathetic disease. Ralph never got upset when one of these cards appeared in the production manager's hands. He just shook his head while I turned away and bit my knuckles.

For a very short period of time Sam was in charge of ordering animal food. As long as it didn't get out of hand, Ralph let him have his devious moment. His favorite deceitful game was to say coyotes had invaded the chicken coop and wiped out either the rabbits or the hens. He had in fact taken them home, to bring them back in a day or so and sell them back to his son at a higher rate.

He never really did anything to hurt anyone; he just liked having one up on us all, and for all his devilishness he had a likeable way. He wasn't a mean man, just a little out of step, and we all knew him for what he was and put up with it for Ralph's sake. But I was discovering he wasn't too good in emergencies.

"Sam! Snap out of it. If you don't get me to the hospital, you'll have to deliver the baby yourself." That brought him around. "Ralph's on his way home, but it's going to be too late! We've got to go now." And I propped up my stomach with my hands. Sam said he'd take care of everything. He left the house. (Somehow I had the feeling he'd never come back.) And I ran in and shaved my legs.

It wasn't long before I heard the familiar rumble of Bessie's engine chugging "I think I can" up the hill to my cabin. I had forgotten about dear Bessie. I ran out, overnight case in hand. Prince started to howl and bark and scratch at the window till I thought he'd break through. I ran back and let him come with us. "Faithful Achates" was to be part of the waiting room.

"Oh Bess, I've never been so glad to see anyone in my whole life." I bent down and kissed her hood. She was missing her fender; she was also missing the floorboard and the windshield, but Bess had come to the rescue. (The knights in shining armor were never around when you needed them.) Prince and I climbed in, and we were off. I couldn't help watching the asphalt below my feet as it passed at thirty-five miles per hour. I kept wondering, what if I fell off and had to run to keep up with the practically brakeless Bess? I wiped the moths off my face. Prince lay in what was remaining of my lap, mainly my left leg. I had to hold his paw to keep him calm. Sam was a nervous wreck behind the wheel, and my aches were so close together now that I didn't dare drive. Twenty miles from home we passed Ralph, speeding the other way toward the ranch. Bess didn't have a horn; so I watched behind me as his lights came to a close with the curve of the highway. "Oh well, the way we're moving, he'll probably still beat us to the hospital." Sam wouldn't answer. Prince wouldn't stop staring at me, and every time I gasped for breath, he whined, and I had to reassure him by pressing his paw; yet his concern filled my heart. He was a comfort.

We must have looked like the Oakies from Fanokie at their worst. Bessie's flatbed was loaded with crates, rakes, and a pile of elephant bedding blowing its trail-marking way down the asphalt highway. "I think I can, I think I can" Bess hummed as we ran one red light after another in Van Nuys. Sam was mesmerized and didn't say one word the whole trip. I'm sure he felt that as long as he didn't ask me how I was doing we'd make it in time. And so we did (barely). Tana was born fifteen minutes later. I left early in the morning, refusing to stay any longer than necessary. There were people screaming in that place! I walked out, but they had to carry Bess home on a U-haul stretcher.

–Yes, I was blessed to have a friend like dear Bess.

Tana's godfather was a lion.

The land shook beneath the thunder of jungle roars from the canyon on the night Tana was born. The girl cub was our only child (we now had one of everything), born to a life where no human babies had previously existed. She grew as if from the pages of Rudyard Kipling and thrived under the loving care of Kasan, the great grey timber wolf, assigned as protector of the crib that housed her new life.

Kasan, raised by a lioness when his mother abandoned him, was now raising a "man-cub" who would one day run with his own.

My baby shower had not been the usual sort. Nor was Tana's childhood like any other. Her godfather was a lion, her Nanny a chimp. Her play yard was equipped with an elephant, whom she used as other children would a jungle gym. A snake named Excalibur was her rubber sword, and occasionally (when Mommy wasn't looking) posed not only as a vine, but as a jump rope. "Tana, Excalibur is *not* a swing!" All the apes were potty trained together, and they wore Tana's hand-me-down things. Sultan, Serang, Patty, and all the other tigers were among the playmates of a special child. There was no mean Shere Khan to hide from here, and she was raised in the same way all our animals were. Because of this, Tana was blessed as no human I have ever known.

A big family of fur and feathers loved her as one of their own. And so it was that she grew and thrived among the animals. Mowgli took shape in the form of a real-live, beautiful glow of a golden girl in a homemade jungle, who had little to do with her own kind until she began school.

Tana's play yard came equipped with an elephant.

The playmate of a special child.

Tana and Zamba.

She was raised with all our other animals.

Tana and Excalibur.

Hannibal and Tana on a summer day.

8
Amber Eyes

And when God looked back with pain,
His eyes, moist and wise, were the color of amber
And crowned upon his head was a silver lion's mane.

–R.D.H.

1

People who have a real concern for their particular science discover they want others to benefit from their experience. In order to perpetuate their knowledge, they cease a good deal of field activity to develop interested minds. As mentors, they begin to teach.

I can guarantee that if you were to walk one day with Ralph among his animals, your life would be touched, as mine has been. The character of our ranch makes us proud to show visitors around. We were proud of the new concept we had developed in animal training that had made us unique. We found happiness in our accomplishments and were delighted to explain to those interested all we could about our unusual life with animals. Half man, half animal, we were all friends evolving through life. Elevated through our work, we found contentment–a condition many seek. So many people came to visit, and so many wanted to know more.

That is why we opened the Wild Animal Training School, hoping we could give people a better understanding of exotic animals. We became teachers, and the concept grew. Ralph opened those first classes with this introduction.

> For the greater part of my life I have carried forward the work of training animals for the movies and television industry. I have done so with the firm faith that what I was doing was releasing animal kind from the shackles of cruelty and disregard shown by many others in the field. Making my business a qualified and highly regarded profession has occupied thirty years of my life. As an observer for many years, I tossed about among the hierarchy of established trainers, who reigned unquestioned because there was no comparison for judging their methods. Their prerogative determined the laws of the working animal industry, and their rules dominated the way things were to be done. Since no one knew of another way, everyone was submissive to their lethal know-how. When, in the '50s, I began to question their methods in open retaliation, I was nearly overcome by their hostile aggression. No one fought the battle with me. Certainly others were fighting against cruelty to animals, but they did not have an alternate method for training the working animal; so I stood alone on the front line–a one-man defense–fighting energetically, but still alone, hoping to clear the way for others to follow.
>
> My theories have developed through practical application and have endured through an evolutionary survival of the fittest. Some people have tried to discredit my work, no doubt; for none of us has grown without making his share of mistakes. And in the process of my development I have certainly been subject to review. In a new field, the way is not always clear for us to perform without controversy, but in my investigations I have tried to direct captive animal behavior by fostering man-animal relationships.
>
> So for as many times as I was wrong, I hope there were as many when I was right. Only time can qualify my work, for I have neither hidden my research nor worked underground. Though the film industry has suffered many growing pains of its own, and though I have walked off many a set rather than subject my animals to the inexcusable demands of a director who had complete disregard for life in any form, it was this

medium that offered me the opportunity and training ground to prove my methods. And though many people consider Hollywood the wrong side of the tracks–lacking in credibility, I am in debt to her and feel only gratitude. The great majority of productions I have been involved with were run by concerned, considerate men, who always consulted me on the feasibility of the scene before the animal was allowed to perform on the set.

I firmly believe a person must have an unshakable belief in his or her convictions, not that he should consider himself above question, but his convictions should be durable so that they will outlast those of people who fall by the wayside because of doubt. Since society is man-made, man must direct learning through his experience and must establish fair laws based on knowledge for all life in society's unnatural state.

My great desire now is to organize my findings in the form of an educational institution, where conclusions can be drawn and elaborated upon with the goal of furthering this natural science. It will be a school where animals may benefit through the growth of man's understanding. We look forward to a time when the public will be well informed on the subject. Man is but a small part of the population of this globe, and it is my opinion that unless we begin to comprehend life and her similarities, our moment may have passed, and an afterthought will have come too late.

If we stand together now, unified as a forest, yet apart as trees protecting a sanctuary, we may emerge strengthened from the shadow of ignorance into a society where art is recording life as life diminishes–a society that should be making a great effort to keep life alive, fighting for those who have no rights.

Together, through inquisitive minds, we may understand from the animals a way to save some of tomorrow.

2

In 1964 we took a partner for the first time, a man who hoped to establish himself as the new Disney. He had a production company, and we had the ranch and the animals; so we merged and formed a financially

Ivan Tors, Walt Disney, Ralph, Serang, and Sultan. (© 1965 Walt Disney Productions.)

Daktari: *Cheryl Miller, Marshall Thompson, Judy, Prince, and Serang.* (With permission of MGM Studios.)

Clint Howard and Gentle Ben.

successful relationship. During the four-year association, we developed and filmed three television series: *Daktari, Gentle Ben, Cowboy in Africa,* and a dozen or more motion pictures, all of which were held together by superior animal work (if I do say so myself). Ralph not only provided the animals for these and other productions, but he also wore many other hats,

Clarence, the crosseyed Lion. (With permission of MGM Studios.)

occasionally serving as writer, producer, and director. With the airing of *Daktari,* the American public was for the first time to see wild animals intermingling as individual personalities, and a new regard for them was established. Hunters, among others, wrote to us and swore they'd never shoot again. New laws were written to protect the wild animals from being treated inhumanely, and Ralph and I were beseiged with requests to rescue animals from the wild who were threatened by death.

During this interval of our growth, we were in tune with the universe. One might say of us that we saw the world through a lion's eye and possessed at times a savage humor. The ranch began to change in appearance as though on a spree of self-indulgence. A sound stage now stood where once had been empty space, and next to the imposing theatrical structure we built the offices that held such great promise.

Inspired by so much orderly confusion, the mechanics' shop now hung out its field degree and began taking skilled analyses of ailing machines and so practiced makeshift medicine.

On a rise not twenty feet away, my circular Junior Zoo had been erected. This was where the babies, having outgrown the nursery, were now living in a kind of animal suburbia in tract homes. In the center of the nursery rose a 12-x-30-foot arena, where every day the young ones

Fluffy. *Tony Randall and Shirley Jones.* (Courtesy of Universal Pictures.)

were turned out to play with one another. Here the elevation of animal kind began to shape; this was where the year-old animals began their formal education of etiquette and refinement. Their code of social behavior and prescribed courtesy became a matter of routine. The training school was held here.

Next to the circle an obstacle course had been laid out, where patient people and confident animals began their educational process. Whenever there was free time, everyone came to audit these classes, and it was a pleasant thing to see the edification and knowledge resulting from such a place. It was quite possibly the nation's most unusual institution for advanced learning. There were tunnels to crawl through, shredded canvas doors to walk through, stationary scarecrows to jump upon, barrels to zigzag around, piles of trees to walk across, walls of mirrors to confront, stairs to climb, ditches to jump, wobbly bridges to cross, teeter-totters to ride, recordings of loud city noises to pass, old trucks to hop in, racing lanes to run in, and almost every kind of thing one could think of to introduce the animals to the outside world.

It appeared more of a playground than a training ground, but by the time an animal had grown accustomed to such chaos, the recipient of practical experience, he was able to conquer most of the city's obstacles with confidence. The obstacle course was a place none of us ever really graduated from; we used it whenever possible for a brief refresher class. Going to that school was so much fun, we have remained students.

Directly across from the Junior Zoo was the wild section, which quartered many animals deficient in common sense. Some maintained a nasty, self-destructive negativity, which they reinforced with an obsession and tried to exert excruciating pain upon whomever they could entice within reach. They were bad guys of ill-bred temper and genetic deficiency. The transmission of such heredity results sometimes in criminals with regressive tendencies.

The reproducing animals were also located here in a maternity row, and we practiced a highly successful program of selective breeding, producing rather a master race of well-bred animals. We chose for each male of special quality a female of equal combination, and their offspring usually resulted in babies of exceptional character. It became an early emancipation of certain genetic strengths.

Next to the nursery, in the center of the park on one side was a primate house, in which also resided the indoor plants used sometimes as dressing on the sets. The combination of chattering inhabitants and green hanging foliage gave one the feeling of a sheltered forest. On the nursery's far side was a cafeteria, quite large and comfortable, whose walls were decorated with posters of films we had made, and movie-star animals were featured in photos above the executive tables. Meals were carried from a stainless steel service counter to checkered table tops, and crews from all over the world who came to film at Africa USA were fed here. The view from these windows was of several acres of pens filled with lions and my beautiful, sensitive cheetah.

To the left of these enclosures was the *Daktari* set, and behind the rhino barn was a man-made rain forest, thick with dense, broad-leafed plants and tall, relocated trees. Artificial vines draped from above, often twisting about live branches. Sunlight had a difficult time filtering through the overhead camouflage net. A stream ran blindly through this tropical, flourishing jungle, where there was seldom a dry season because of controlled spray mists dropping hundreds of inches of rainfall each year. The alligators thrived in this steaming swamp region.

To the left and east were the stables, and along the railroad tracks sat Beverly Hills, where the famous of our clan found dwellings away from the thousand animals who now lived here—the largest privately owned animal facility in the world.

At Beaver Dam, near the lake where the elephants took their daily bath, planted palms and pampas grass and philodendrons grew, and the trails wandered in red earth. Atop the largest oak perched the tree house in which Tarzan had lived, and past this lovely spot stretched a sandy coastal plain, where African villages and sets of all nomadic kinds vanished in migrating season. At that property's end stood a tiny little restaurant, which we kept open to the public for undernourished passersby. The ranch stretched for nearly one mile along Soledad Canyon, and the acquisition of new property gave us a good deal more to think about. But no two people could have enjoyed a place more than Ralph and I did our Africa USA. We groomed each speck of dirt as we would fields of unnamed seeds and watched our harvest emerge.

When time allowed, I would sit beneath a summer California sun at the edge of Swan Lake, skimming rocks and contemplating for long periods things that made me feel in good company. Sometimes, propped against the corpse of a dead tree with Sultan beside me, I became a diligent student of nature. Independent sounds mixed and blended, deepening my perception. These moments for me were solitary satisfaction; I could not ask for better. My tiger, the lake, our ranch—these created in me

With Sultan beside me, I became a diligent student of nature.

a poetic intensity. Here, all things seemed to be sincerely represented. I had come to love everything in my life. Years of honest searching had brought me to a promised land. I was deeply satisfied in such a place, where I seldom identified with the human race.

The elephants cleared fallen logs and other natural debris from the park, carrying a heavy load in harness. They cleared the way as elephants do in their place of origin, as their fathers and mothers did before them in the grand teak forest—pulling and straining at the dead, rooted trees, coupling their great physical power with an intense concentration. Their tasks demanded varying degrees of effort, ability to perform effectively, and extreme control. They exerted an enormous influence over us; we envied such strength. But because we were the regulating influence over that strength, we felt a certain standard for comparison, as though operating a mighty machine. It was a personal satisfaction to speak to such giants and have them respond to our small voices. And by having such friends, we felt rather like monarchs reigning over a powerful nation.

At times in the lake at Beaver Dam, we had grand battles atop our elephants. With brooms we toppled each other into the water far below. Then with their trunks our mighty steeds carefully replaced us behind their heads, and the battle resumed; now the pachyderms got into it, and trumpets blew as heads came down and bodies collapsed, giving inward suddenly and folding completely as they became ships capable of operating submerged. Bubbles blew all around us as the submarines went below the surface of the water, and when the optical instrument came up for observation and a breath of air, a wave of muddy water came down on us with such force that we were again capsized.

And now the periscope turned before our very eyes into a monster serpent, lashing about as uncontrolled as a fire hose gone berserk, and we swam screaming for the shore with the thing coming at us. A boulder erupted from the primeval waters, and Modoc came shaking with humor to my side, and she and her regiment rolled about in the fine dust with such belly laughs that one would have thought them fit for a mad animal house. When they resumed their composure, they rearranged the dirt and threw it in a storm into the air, covering us all so that we sneezed until we nearly blew up.

After such a dusty soiling, we felt an undisputed obligation to cleanliness, and the herd of us ran from the murky banks, some sliding all the way to the waters below. And now with brooms we scrubbed the mountainous walls.

Sometimes we ran—Ralph, Modoc, Mardji, and I—side by side, and one would have thought the earth about to crack from such a weight. When the elephants took us for long rides in the hills and we perched, examining all things, I wondered at the thoughts of the creatures we passed—small, wild inhabitants of the region. Were we for them fearful lumbering monsters dating back four million years, for ten thousand years now extinct? How was this restoration possible? They would, on occasion, review our presence with a critical eye and report on us to one another. Some introduced themselves directly as we aroused their interest in a country other than their own, and others stood perfectly still, frozen as though preserved in ice so that we would hardly take notice of them. Some ran below the earth, occasionally popping up from a hole systematically to inquire of such an unbelievable sight. Some, distrusting their senses, gazed on us in disbelief. Others were bored by such an infringement and would not admit to our existence; blasé, they refused to move from our path, and we would swerve to miss such opposition.

3

As the head of the training school, every day I rediscovered the excitement of my existence and for the most part remained unalarmed. But I was often put to the test. Maintaining my femininity where I was the only woman was something that didn't really concern me but occasionally required that I remind others.

The perils of Pauline pale when compared with my life as a stuntwoman and animal behaviorist. I have amazing confidence for one who recalls hoping for a terminal disease rather than appear on a live TV talk show.

It is only natural, for example, for people to be concerned when an eight-hundred-pound bear runs amuck on a sound stage, even though he is supposed to destroy the set. Although I know the situation is controlled, some people have their doubts. One of my early experiences resulted in a brief encounter with the kind of fear that characterizes some men. My friend Bruno was playing his part brilliantly, portraying a demented bear intent on demolishing the interior of a cabin. I waited with his pick-up leash on one side of the set and Ralph on the other.

When the director called "Cut!" Bruno (later to be known as Gentle Ben), a seasoned actor who knew film commands as well as those we had

The perils of Pauline.

taught him, ambled my way to share his between-camera break with a friend. As he lumbered toward me with an air of boredom from the artificial room he had just devastated, he stood on his hind legs, checking out the booms hanging above. A big, husky stagehand stood not far from me, and as I walked toward the bear, the petrified, imposing man grabbed me and shoved me in front of himself. I fought to free myself, and only when

I polluted the atmosphere with a startling word did he release me.

At the school I met many men who had abandoned their former culture to adopt new ways of life. But on the set I met over and over again that same kind of impolite personality that seems to characterize a human condition. Because of many similar incidents, Ralph put his crews in a portable cage, which he had designed to house primitive thinking men while civilized animals were acting. And while they were locked up, he called a gathering of their minds and developed in them a new understanding of animal actors. I have now worked with most of the crews in my town, and this Christian with her religious energy has never been grabbed at again—at least not with self-preservation in mind.

My life as a stuntwoman—which continues to this day—was begun as an unskilled laborer rather than as an artist. It's a wonder I walk about with all my limbs intact. I was never really trained to preserve myself in that element. I was, as they say, an accident just waiting to happen.

A producer walked up to me one day and asked, "Are you a stuntwoman?"

"Well, of course," I replied (the way everyone in my town does).

"Good, then you can coordinate the stunts on my show!"

And that is how my career as a stuntwoman began. Little did that man know the feats he believed me to display with such skill and daring were in fact "the real thing." They weren't stunts at all; they were *accidents,* which fortunately I was able to walk away from.

Inspired by youth but lacking in finesse, I accepted all the stunt work for the two series we were filming at the ranch, *Daktari* and *Cowboy in Africa.* I think the idea of Ralph doubling for women was my chief encouragement, coupled with the unequaled monetary considerations. I developed a rather philosophical nature regarding my fate: quite simply—I was indestructible!

One day Superwoman had to wrestle a lioness. In the script I was to start my run from camera right, and approximately thirty feet in front of camera the cat was to catch me and the fight was to begin. Ralph would release the cat onto me. It was an exhilarating morning. I could take on the world. "Whenever you're ready," the director called.

"Ready when you are, C.B.," I replied, assuming position.

"Rolling cameras. All right, Toni, *action*!" And I took off like Babe Didrickson. I heard "Release the cat."

Things were going well on my overland journey, and I was just making my turn to catch Tammy the lioness in mid-air when suddenly we both, as they told me later, simply "vanished from the horizon." I found myself sprawled in a rather deep ditch, staring up dazed and winded at a very astonished lioness, who was sitting on my stomach. "Somehow, Tammy, I don't think things were supposed to go like this. Okay, who put the pit here?" I screamed from the hole, pushing the four hundred pounds off my body and frowning up at the "still" man, who was clicking photos of us as fast as his motor drive would allow. "I find that very annoying!" I complained, sidestepping the bouncing lioness. Tammy was having more fun than my battered body could endure. Having me all to herself, she was leaping all over my head. "Now cut that out!"

I was trying to decide how to apologize when the crew gathered around and, gazing down, began to applaud, and the director called out, "We'll write it in. That was fabulous, Darling–absolutely fabulous. Do you think you could wrestle her down there?"

"Well, what do you think I'm doing?" I answered, plucking gravel from my ivory complexion.

"Now, Darling, don't get angry. Just tell me what you want to do."

Angry? What does he know about angry? I was about to explode, my face aflame. I compromised by asking for $500 more than the initial $500 I had asked for.

"You got it," he said to my astonishment, and again the cameras rolled.

From that stunt on I mapped out my path of action beforehand. I still had many lessons to learn, though some invisible force seemed to make my worst mistakes turn out right.

We had no slim trainers; so when a slim male actor needed a stuntman to do a scene with an animal, guess who became his stuntMAN? You guessed it. I took personal affront at doubling for men. They had to bind my chest with an oversized ace bandage. Sometimes I wore a beard. And once they actually put hair under my arms.

I was tossed from more horses before camera than I care to remember, biting more dust than even my long-dead relative, Johnny Ringo. I rode bucking rhinos, dodged charging cape buffalo, raced giraffes, and outlived my days of pain. I hurt in places I didn't know could hurt.

Nevertheless, the stunt woman in me lives on, and with the lapse of time and an abundance of experience, I know better and keep telling myself to retire; but every time the gauntlet is thrown before me, my blood races, the thrill overcomes me, and I can't escape my love of working with the animals–the challenge of that next feat.

And the world wrote about us then. (Photo by Richard Hewett.)

4

The ranch had a quality about it that aroused the curiosity of many people, and visitors from all over the world came to investigate—some with supportive intention and others expressing suspicion. We were probably at our peak during those years from '64 through '68, but we had no time to think about success then. Neither had we time to criticize, analyze, or evaluate our days, so caught up were we in keeping the great machine going. But the inspiration of growth as dreams came true was a stimulus to mutual enthusiasm in us all. For a while everyone seemed to whistle the same little tune, and there was a strong unity in all our lives. The world wrote about us then.

Ralph does not dash about the continent stepping on people's faces and spitting in their eyes. When he speaks, it is with authority, and he is not greedy in his desire to be right. For a while we had been under severe scrutiny by many zoo directors, who, in their undeveloped state, dealt with the only facts at their disposal. The anodyne, which in our case was the credibility of our work and the undisputed safety record of our company, eventually relieved us of their prejudiced aggression. The bigot became the skeptic, and with the skeptic we could talk.

The discussion usually concerned the controlled environment in which captive animals (with no hope of returning to the wilderness) lived. And because the zoos are responsible for bringing the exotics into captivity, we believe more of our zoos should develop programs for emotional rehabilitation. (Some of them have indeed accepted and carried out the concept of affection training.) That animals should live for a lifetime in sterile isolation and in a constant state of anxiety seemed horribly cruel. And the men who defended this way of life so rigorously were men who had, for the most part, no empathy for the suffering caused by confinement—of days with nothing to look forward to. Knowing what I do now regarding the psychology and philosophy of animal behaviorism, I sometimes feel sick when I visit such a place—helpless in that I have no say there. The directors of those places, with their varying degrees of judgment, would always reveal a camera when they came to our place, shyly requesting a photo of themselves posing with a tiger or a lion or perhaps sitting atop an African elephant. They could not do this in their own zoos; their animals were too wild.

9
T.R.Helfer's First African Journal

You are free, free to roam this world with me,
until such time as you are really free.

–R.D.H.

1

Daktari and *Cowboy in Africa* were filmed entirely on the compound. The *Daktari* sets were built just next to the hippo pond, which sat at the foot of the cat circle. And although the pilot for *Cowboy,* which was called "Africa Texas Style," was shot in Kenya, the series was filmed at the far end of Beaver Dam just below Chester Higgins's house. And for those four years that we filmed at Africa USA, not a day went by that Old Chester Higgins didn't start his buzz saw at the precise moment the director called for action.

To our surprise the overwhelming popularity of these shows brought fans who went so far as to dig under the fences to get inside our ranch, which was closed to the public. Many hopped a free ride on the freight train and jumped off on the side of the ranch that wasn't fenced, and one fellow actually bailed out of a plane in a parachute. Others camped by the roadside, hoping to see their favorite stars, who had become overnight successes and were instantly treated as American royalty.

Burt Lancaster after death scene in Island of Dr. Moreau.

The Duke (John Wayne) visiting two black leopard cubs. (Photo by Jack Henderson.)

The invasion of the ranch by two studio crews, not to mention every celebrity who crossed the southern California borders, was enough at times to make me run with the screaming meemies from the never-ending social functions.

2

Stevie was leaning back in her chair propped against the rim of the open cafeteria window. I sat across the table from her, and from where I sat I could see Ralph coming toward us, his finger pressed to his lips, signifying "Shhhh." By his side was Zamba. I could hardly keep my face straight. I put my fingers to my temples, lowered my head, and said in a squeaky whisper, "I have a sense of danger close and threatening."

"Toni?"

"You have my undivided attention."

"Have you any idea how strange you have become?"

"Why, what ever are you talking about, Dear?" I quizzed with a fanning of lashes. "Wait, before you answer that—cover your eyes."

"Cover my eyes?"

"Trust me!"

"Trust you? Surely you jest; hast thou forgotten who I am?"

"So as to clarify your identity, cover your eyes."

"Why not?" she huffed. "But let me state here and now that I do not feel optimistic."

"Well, you're only human. Now do as big sister says—cover those big brown eyes while I stunt your growth."

"Excuse me. Did I hear you correctly? While you stunt my growth?"

"Well, someone has to do it."

"Toni, what have you been eating?"

"Why, monkey biscuits, of course. Want one?"

"No thanks, I'm driving." (And she finally covered her eyes.) "Lord, hear my prayer."

I was wringing my hands with anticipation as Zamba jumped up and placed his two front feet on the ledge just next to her and yawned with slow deliberation as I said, "Turn your head to the left and open your eyes—there's a good girl."

"Eee gads! Yikes! How simply ghastly!" She dramatized, putting her hand in Camille fashion to her forehead. "It fouls the senses." And she

Stevie.

fell forward onto the table in an exaggerated faint. She opened one eye and looked up at me. "Are you impressed?"

"Where did I go wrong?" I groaned, pounding the table.

"What can I say? It's all those years of torture that have made me what I am today."

"And what, pray tell, is that?" I dejectedly asked.

"Numb to your foolish pranks." She glowed.

"Well, why don't you just get out the rubber hose. I'll never be able to show my face to the family again." I crooned, "Some fun it's going to be having you around–some fun!"

Ralph poked a zealous face into the window between Zamba and Stevie. Stevie picked her head up from the table, turned to face Ralph, and screamed. "Now that's what I call scary!"

"I've got exciting news," he beamed. Exciting news to Ralph had a wide range of meaning in it. It could be that a duck had dropped its beak and grown lips—or it could mean I was about to be bestowed the honor of wrestling a four-ton elephant. I never knew for sure what was coming next.

He took a deep breath, draped an arm around Zam and announced, "*We're* going to Africa."

"You mean you and Zamba?" I asked, feeling as though this was where I had come in.

"No silly. You and me."

3

On location in Gorongoza, an isolated interior of Portuguese Mozambique, we sat beneath a somber sky in the shade of a subdued yellow fever tree, having just come from the House of Lions, a deserted safari camp, now occupied by two prides consisting of twenty-odd lions. There were no doors on the stucco buildings, and we could see lions asleep in bathtubs, lounging on the devastated remnants of furniture, sprawled idly in doorways, and climbing the spiral stairway leading to the sunroofs, where they bathed lazily in a blazing inferno.

My mind, as it so often does, had taken a vacation, and I was blissfully out of touch with reality, playing in primal fashion with Ralph and our chimp McGuinness, who had come so far to double for Judy. We were filming a second unit for *Daktari* to authenticate portions of the series now filming at our Saugus Ranch. Suddenly we were aware of fourteen Africans surrounding us and staring at us. Chilled at the sight, we leaped to our feet. Mac barked, and Ralph caught him by the hand and lifted him to his arms. Deep shadows moved in on us, and we were as prepared to die as we have ever been—not thrilled, but prepared.

No one said a word.

The leader of the savage-looking crew took a giant step forward, eyes bulging, jaw protruding. (Oh, God, please don't let them be cannibals.) He reached out and attempted to touch the chimp, who, with utter repulsion, barked and hooted at him. The decorated man nearly deposited his spear into his foot as he jumped straight upward, and the other natives screamed in some indistinguishable jibberish, scattered, and ran (along with their fearless leader) back to the bush from whence they had come.

We could not get out of there fast enough, and as we headed for camp I asked, "What do you suppose they would have done if Mac hadn't frightened them away?" There was no supposing. I was sure they would have thrown us into the pot.

Later that evening, after dinner, we left our thatched bungalow, made ourselves comfortable on our walled-in patio, sat back, and began to read. McGuinness had a full basket of goodies to occupy himself with. He busied himself by tearing beads apart, then, with great frustration, tried putting them back together. Suddenly he stood, looked out into the night, jumped back and forth, stomped his feet, barked, and made for Ralph's lap, where he hugged him and made hyperventilating sounds as he placed his mouth over Ralph's chin, then nervously looked back into the dark.

"What do you think he sees out there?" I asked, squeamish.

"He's such a scaredy cat about everything, it's hard to say."

But I felt the discomfort, too. Everything was too quiet.

Guerilla warfare was going on in Mozambique, and I had the feeling that since we were in so remote a place, we may have been marked as the next victims. I picked up a nearby poker to defend the fort when a piece of ceremonial jewelry landed at my feet. I reached down and picked up the necklace, adjusting my eyes to the approaching gloom. I thought I saw movement. My heart was in my throat, and my hand clutched the weapon I was prepared to slay with. A figure emerged from the night. He was very tall, and I could now see hundreds of eyes around him. Everything was black on black. I was about to attack when the man spoke in perfect English with princely dignity, "Excuse me, but is that your child?"

Child? What child? There were no children here. What was he talking about?

"I'm afraid I do not understand you," Ralph acknowledged.

The man pointed directly at McGuinness. "Him, sir, is he your child?"

We looked at one another, not knowing quite what to say. For a chimp Mac was quite handsome. But for a human baby! The thought was frightening, and besides I took it as a personal insult that someone could think I had given birth to *that*!

"I am a Christian," the man said slowly, "but these people are not," at which dozens of Africans cautiously stepped from the night. "Word of this boy has spread through the villages, and they have come to honor him. May they come forward and see him?"

They seemed so serious, and we certainly didn't want to offend them. Ralph whispered to me, "They've never seen a chimp in this part of the world, and I can imagine what a shock it must be to them." To the natives Ralph said, "Yes of course they may, but don't let them come any closer than the edge of the patio."

Ralph attempted to explain to the man what Mac was and where he was from. The man said "Yes, yes," and nodded his head, but it didn't seem to penetrate.

Some people had gifts, which they left on the wall. Some held hands as they came closer. Some laughed with embarrassment, while others seemed to chant a prayer. A witch doctor of sorts came shaking some rattles. It was all quite ceremonial, and we did not say a word but took it all in. One silent hour later, a long line continued. I looked at the sky. There was no bright star they might have followed–no camels bearing kings. And although Ralph toyed a bit with carpentry, I was definitely not Mary.

4

The herds were migrating while we were in Mozambique. Ralph, Mac, and I, caught up in the spell, stepped out onto the plains one early dawning and took our place among the dazed beasts. We walked and walked within the center of the herd, and not one took notice of us. We were so close one would think I could reach out and touch the sides of zebra. And if he had wanted to, a wildebeest had but to turn and impale me; but not one animal recognized our presence. McGuinness, quite self-contained, rode upon Ralph's shoulders in alert attention, occasionally looking to the left, then to the right, at the curious procession. As we walked, others joined in along the way, and for nearly two miles we stayed with the noisy mass, then slowly stepped to the side as the parade passed us by. When we were alone and the swarm was long out of sight, I asked Ralph. "Why did we do that?"

And he replied, "I haven't the vaguest idea." I found that a bit startling. "We seemed to be caught up in something universal, as though we were swept by primitive instinct into some past natural state."

"Have you ever done that before?"

"No. It was like being part of a vast energy field, where to go against the powerful impulse would be violating a natural law. We were for a time answering a primitive call." I was beginning to think we'd been alone too long.

McGuinness used to derive great pleasure from tormenting the Africans, having discovered that if he charged and stomped like Rumplestiltskin, he would send the fearful blacks streaking into the dark bush. Fortunately for them, his reign of terror lasted only a short while.

One exploratory afternoon we three apes drove deep into the inland part of Gorongoza, accompanied by a large basket filled to the brim with fruit and two Africans–one to drive and one to see that we didn't get lost. We came to a spot where Eden intervened, briefly modifying the two worlds. Here in this splendid place of reviving density we stopped in the shade of a yellow acacia, receptive of the surrounding magnificence, not as intruders but as spellbound captives. Inspired as always, Ralph began to philosophize.

"Do you realize that for millions of years the animals have ruled this continent and the black man has respected their laws? It was only at the advent of the conquering white that their harmony was forever disturbed–interrupted by social distinction–caste immobilization–singular persecution. I tell you, Toni, when I look back at that city, Johannesburg, I feel nauseated and ashamed." And for half an hour we discussed injustice and ruthless cruelty, stopping to eat a banana or a slice of melon.

Then Ralph, having temporarily relieved himself of a mental burden, stood up and with extraordinary vigor took a long stretch as a lion might do, then fell into a series of pushups. Matter-of-factly he announced while pulsing, "Did you know the Mamba possesses an extremely violent neurotoxin that paralyzes the muscles and impedes respiration, causing a terrible death by asphyxiation?"

"No Ralph, you don't say. That's news to me. Is that my trivia lesson for today?"

"I only mention it, Toni, because I'm told there are many in this particular region."

"*What*!" I jumped up. "Ralph, why do you do that to me? You know that makes me crazy."

"Yeah, I know," he said with an evil look in his eyes.

5

From the sitting room at the club where I record in my journal, I see the clouds still and hanging on the mountain; smoke from Kikuyu huts stretches skyward in an unbroken path, leaving the earth far behind. I feel the eyes of death all around me. Mounted on the walls are the heads of what once lived in the heart of Africa, once alive and free to wander these green hills, free to hunt under this vast span of sky. Nature's children, never to play again–here, hanging from nails on cold, white walls, death's vacant stare. I want to tear them from this place, to bury them and liberate their souls, but if I were to do so, "they" would only replace them with the new dead. So I choose to leave them as a constant and grotesque reminder of man's lack of appreciation and his injustice toward free spirits. Some among us can't bear to see anything free.

Kenya is my first encounter with the tourist, and from a rack I see machines dangling: lenses, binoculars, cameras, tape recorders, light meters—all hanging in excess weight from soft, khaki-clad bodies. New clothes, new shoes—never worn and with no stories to tell. Everyone is the same. Stamped, signed, sealed, and delivered in Volkswagen buses.

"Photo, photo, only two shillings." "I want to see a lion. You promised we'd see a lion! Now you show me a lion!" "I've seen a lion, for goodness' sake, I don't want to see another!" Photo, photo, click, click. Hurry hurry, racing racing. "Oh no—I lost my film!" Proof vanished, you were never really there.

6

Dawn on an African day is a rich legacy, pleasing to the senses. A great, red sun peeks a fiery head above a vacant, vast horizon; and in the light filling the land, a distant vapor to the east inches its way toward the fog-capped mountain island in the desert known as Marsabit. Sweeping life comes near, and the faint sound of bellowing. A grey wall of elephants is now visible, ears directing a current of self-made breeze, legs so strong they pack the earth. Moving at a rapid speed, they approach the mountain's edge and veer into the sanction of hanging mist.

As we found the herd, they were grazing in a lovely patch of green, wet glades, left damp by a favorable rain. But absent from the compact mass was Ahamed, the mighty legend of Marsabit. Ahamed had seen fit to conceal himself in the cloak of the enshrouded mountain. He was not to be found, that day or yesterday or two yesterdays ago, and our vagrant hopes kept on the go in random fashion. We drove through dense forest where heavy-leafed plants succumbed to the thick air. In places the deep grass was so high we were unable to see, and we had to maneuver blind, for we were lost. Having searched for many hours, we stopped at the shores of Lake Paradise. The Kudu remained silent on the opposite bank, while birds flew above our heads, questioning our intentions. About the water's edge steamed piles of pungent dung. Had the mammoth recently been here? The remains of a camp above the lake, a spot made famous by Osa and Martin Johnson, sat on a small hill overlooking the great expanse of crystal waters. Ralph and I walked through what once had been a flourishing vegetable patch and gardens fragrant with flowers. We stood on the empty ground where the memory of them has lived on, and at this haunting place we were filled, enough to resurrect a vivid image of the way it once had been.

It had indeed been a very long three days. It seemed that we had covered every inch of this remote forested mountain, rising so abruptly from the otherwise barren wilderness. Although we were sad at having been unable to locate Ahamed, we were fascinated at the fertile site.

Dusk and the darker stage of twilight accompanied us as we drove homeward on the deserted hand-carved road. Our mood was somber, punctuated by our own disheartening and diminutive remarks. We had come many hundreds of miles through waterless country, on single-lane dirt roads; we had been baked by a merciless, dry heat in the desert mosaic, hopeless in our pursuit.

Suddenly, around a very wide bend, perhaps ten minutes past the oasis, an indistinct grandeur loomed, so large it was a commanding structure. The threatening form statued on a rise, and as we grew close, the shape appeared to move.

We were spectators caught in the drama of an African sunset. There before us, silhouetted against the raw light of the horizon, stood the legendary Ahamed, rocking with the ages. And from this natural pulpit he gazed down benevolently upon us. We seemed in the presence of an ancient wonder. I don't recall ever having witnessed such exaltation prior to this moment.

Some people worship mountains and live at the foot of such an altar. Somehow, size denotes strength, wisdom, and timelessness. But other people tear down the hills, and men shoot elephants in defiance of God.

Here we stood, so lightheaded as to understand the world, and so caught up as to find no words to express it. From ten feet away we viewed him, and when he spoke, his voice was that of thunder and his breath the wind, and ivory tusks gleamed and cast upon that land long-forgotten shadows.

Oh, Modoc, if you could but meet Ahamed, what a single moment it would be, perhaps the greatest hour of all the hours of such a long life. And you two wise old souls could roam the plains of Marsabit and rule in this place called Paradise.

10
The Wild Life

Yet taught by time, my heart has learned to glow
For others' good, and melt at others' woe.

–The Odyssey, *Book 18*

I suppose the worst accident I ever witnessed happened during those early years of public animal-park popularity.

In the northern part of California a run-down amusement center specialized in various types of marine shows. The park was located on the very edge of the bay and, along with its reputation, was sinking at an astonishing rate, shriveling like a drying apple, huddled on a small island yearly growing smaller by its gradual descent into the miles of salt water.

Through a rather involved business negotiation and with enormous enthusiasm, the new owners of Marineworld offered us a proposal that was difficult to refuse. These gentlemen believed so completely in Ralph's unique approach with animals that they were willing to risk millions if he would join them in their new venture. It was a marvelous opportunity for us; not only was the challenge of reviving the ailing sea monster exciting, but for the first time we could present to the public the eccentricities of our life. We were given carte blanche–complete creative control to design a new leisure-time theme.

Visually the park was an eyesore. It rather resembled a bog of soft, white, waterlogged sand–landscaped with the only thing one could grow in such a place, iceplant. For nearly six months of what seemed like never-ending mental and physical exertion we spent twenty-hour days, seven days a week, tearing down the old and putting up the new–designing, building, supervising, criticizing, analyzing, and revising what we had created. The result was an exciting blend of spectacular entertainment and a theme of conservation and education, for we had relocated over 200 of our menagerie, merging land and sea animals for the first time.

Looking back now, it's hard to believe that we put so much moving force into decorating that arid spot. Yet an enormous amount of energy and capital went into the relocation of tons of dirt, hundreds of full-grown trees, and miles of lush turf that rolled out like a carpet to transform acres of parched wasteland into glorious, rich, green glades. Every vacant piece of forgotten ground was covered with flowers or creepers and blooming things. Paths of redwood bark wound around the three islands. The old buildings were reinforced and given a whitewash, on which I painted black silhouettes of giraffe and kudu, of eland and elephant. The grey concrete patio at the park's entrance was painted ocher, and in its center a huge olive tree spread its branches over a purring cheetah resting in the lap of a pretty girl. Around the plaza, llamas, lion and tiger cubs, and young animals of all kinds marched with their handlers. In this controlled environment the public was allowed to ask questions and touch the animals and simply enjoy a moment of pleasure. The round, thatched-roofed restaurant with its white and black silhouettes, the ocher plaza, the olive trees, the green, *green* grass, and the parade of happy animals had a dazzling effect on visitors. The islands were connected by arched bridges, under whose curved structures rubber rafts sailed to where the animals ran free.

But rather than bore you with a full account of that body of land, recalled from this remote viewing, let me take you back through memorabilia to one of our moments of illuminating irony–to an unpleasant episode.

The "jungle theatre" was to be our most artful expression, where behavioral exercises, the likes of which our times have never seen, could be regulated for exhibit to the public at specified times. Here we would finally be able to demonstrate to an audience of 2,500 people the amazing potential and wide variety of natural, reinforced, trained animal behaviors. That first season the theatre opened with one of our trainers dressed as

George, the only known "saddle-broke" giraffe.

Dr. Doolittle, riding on George, the very first saddle-broke giraffe.

Following George and the trainer's ten-minute recital of all you've ever wanted to know about giraffes but were too short to ask, a herd of accomplished teenaged African elephants lumbered onto the astroturf. Until then the African myth had spread like wildfire: Those elephants are too dangerous, too nervous, unworkable, unreliable, untrainable, unsafe. You take all that away and what have you left? Misunderstanding and other points of ignorant abuse.

Before we could afford an African elephant, when a script called for one, we dressed an Indian elephant as an African, attaching false ears over its own and clipping long fiberglass tusks to its short ones. I still laugh when I see one of our old *Daktari* replays. Old Mo, looking somewhat like Droopy Dog, tiptoes quietly into the background set as an extra, trying desperately to hide her identity so that no one in the family will recognize her. When she is almost out of range of the camera, behind the "Whamuru" compound, much to her surprise, her ear on the camera side falls off. And rather than continue on, she scoops it up with her trunk and begins waving it around as though fanning herself nonchalantly in the equatorial tropical setting.

After the elephants came a brief display of wrestling lions and leopards, then flying birds of prey, then chimpanzees coaching their trainers in gymnastics skills, and a wide variety of spectacular entertainment in which everyone was having fun. I thought it the greatest display of man-animal accomplishment ever presented in the United States.

The crowd wanted more thrills and chills. It seems the majority among us are bored spectators who need an emotional outlet for their discontented appetites. They want more nerve-tingling excitement for their money. And so to amuse the mob—a neurotic society where human sacrifice has become a national pastime—into the arena of the circular coliseum was imported a special feature to which we were irately opposed—a wild tiger act.

It is ironic that in 1969 we had been approached by a group of wealthy Italian businessmen who wanted us to restage in Rome the spectacular "Games." They offered to build a massive amphitheater on the city's edge if we would recreate that portion of history that had made Rome famous. To revive the horrors of the Coliseum would be equal to restaging the mass executions and atrocities that took place at Dachau. Somehow to consider supplying this form of inciting fascination in a structure filled to capacity with the deranged seemed to constitute a mockery of the martyrdom of those early Christians who saw fit to die for what they believed in. The project never materialized.

Ada Smya was a buxom blond who had defected from her native Poland in the mid-sixties. Through those wavering times she and her husband had been employed by Ringling and several off-beat circuses. She was from the old European school of trainers who looked upon perfectionism in a performance as a peer pleaser. Their performances were geared to satisfy and gratify their authoritative, strict instructors, whose careers often depended upon the success of their pupils' showcases. Conversation with Ada—because of our lack of knowledge of the Polish language—was one of the great miscommunications of our time. If any of us were to get a point across to her, it often took an enormous amount of explanation coupled with sign language and visual effects. Our little talks often sounded like a bad interpretation by an African Masai portraying an American Indian on the Hollywood screen or a new-world explanation of the ramblings of the infamous Piltdown Man. "You–me–go–eat–food–talk." Although we came from different worlds, we were friends, and I always hoped one day we could sit and have the talk we tried so desperately to have.

Ada was as good as arena trainers come; her act was polished, tight, and precise. Her cats went through their paces with little coaxing. In most cases they were way ahead of Ada, anticipating their turn before she gave them their cue. The tigers were not overly disciplined, nor were they lacking in control, but during performance you could feel the intense building of emotion. On occasion one cat would slug another, and a round of ocher fury would have at it until Ada, with her training crop and buggy whip, would break up the match, at the same time watching to see that the other rogues were minding their manners.

I had deep-rooted convictions regarding this type of training, believing it to be a negative approach. Nevertheless I admired Ada: although she had faith in her schooling, she was open to suggestion. So a kind of between-the-two-worlds combination of training came about.

Thirty minutes before a Sunday noontime show, I came down with chills and a kind of nauseating anticipation. I sat down on the brown couch in the office before I fell down. Ralph looked up from his mound of papers and asked, "What's wrong? You look as though you've just seen a ghost."

"I have a feeling something terrible is going to happen. I know that sounds silly.... What a preposterous thing for me to say. But I'm painfully aware of some danger all around me."

"Toni, I think you should lie down; you've been working far too hard lately, too many hours and too little rest. I want you to take the remainder of the day off, do you hear? You go home and relax."

"Relax? Easy for you to say; you've never relaxed a day in your life, and furthermore, you workaholic, don't patronize me. Let's just forget I ever brought it up. I know how ridiculous it sounds. I'll meet you at the theater for the twelve o'clock show. Would you like me to get you something to eat from the restaurant?"

"No thanks, I've got some goodies in my desk." He pulled forth an apple and some sunflower seeds and threw me a twinkling, nose-creasing smile. "Now go on, get. I've got work to do. And here," tossing me a polished apple, "take your crystal ball with you."

The Sunday crowds wouldn't appear until after church, so the early show was generally seen by a small, intimate audience of three-to-four hundred spectators. Starting at 2:00 a shoving, pushing aggressive migration of 2,500 visitors, their hands filled with popcorn, hog dogs, candy, and soda, advanced upon us, ready to do justice to that old American tradition—weekend with the family.

The tiger act opened the show. I took my seat out front next to the stage entrance where I could clock each display and jot down suggestions where improvement was needed. It made me feel terribly important sitting among the patrons with my little pencil and pad, separating me immediately from them. And because occasionally I do seem to need that surge of self-esteem, I sat out there a lot.

Ada went through her training session with her usual display of self-assurance, loving every minute of it, while I enjoyed my position as critic. When she was through and leaving the arena, the roustabouts began to dismantle the arena and break apart the tunnel of cages, clearing the stage for the next performer. The cages connected together by fastening a U-bar to the adjoining extremities. On either side of the ten enclosures was a door that opened with a cabled ring. When the door was raised, the ring was clipped across on the opposite cage; so the door cables were held open in criss-cross fashion. When the act was finished, the tigers were released one at a time through the tunnel, and each cage door was closed. One by one the tunnels came apart, and the cages were rolled away with one tiger in each of them.

I was unaware that Ralph had just appeared on the scene, standing just behind the curtain, not twenty feet away from where I sat. Not until I heard him cry out above the overscoring finale music did I know he was there.

"You're pulling the cable the wrong way! That opens the doors! . . . *Good grief! The tigers are loose! Get the* CO_2*! Ada! Where's Ada?"*

One of the backstage helpers had called in sick just before show time, and in a frantic search to replace him a member of the staff had recruited a nineteen-year-old employee from the park who had never worked behind the scenes and was completely unschooled for this temporary, dangerous position. In disconnecting the cages, he had pulled the cable the wrong way, releasing two of Ada's cats, who savagely attacked him. I leaped the guard rail with a single bound. Some among the audience stood. Others were unaware of what was happening. Muffled laughter and bewildered shouts were heard as a few of them reacted by running for the nearest exit, not even knowing why, assuming panic, while others remained perplexed but took their seats as the unknowing ushers instructed them to do.

When I bolted through the curtain, clipboard in hand (as though it were a protective shield), I didn't know quite what to expect. Raja, the largest and meanest tiger in the act, stood not ten feet away, his four fangs sunk deep into the muscles at the base of the neck of the young man. A second cat's teeth were imbedded in his leg. Together they were dragging the horrified victim to a secluded spot, where they could eat him as they would any other piece of meat.

The words flashed inside my brain like a ticker-tape news brief: "Aren't you afraid?" I must have been asked that a hundred thousand times in the past, and my answer was always "It's not so much a matter of fear as it is a question of respect. I have a very healthy respect for wild animals."

But as I stood stock still for that few seconds, there wasn't a doubt in my mind. I was *petrified*! My heart thundered away, and the roaring of my blood blocked out the hysteria going on around me. Among the subdued sounds I heard "Keep the show going; get the chimps out there!"

Ada's tigers were not like my Serang or Sultan, and while Ada loved them, they hated everyone else. Coming face to face with one outside its cage was worse than walking into one in the wild; these tigers knew the score and had no fear of man.

My mind slowed down, allowing the hideous scene to pass before me in slow motion. For those few eternal seconds I imagined myself in the boy's place. He was limp, dangling weightless between the massive fangs. His eyes were half-cast, and blood was pouring down his chest. His pitiful, small voice said, "Someone help me" as Raja tried to drag him behind the backstage cages. The second tiger tried to tear his leg off. He held the body tightly with his claws as the other pulled, ripping and tearing flesh in the vicious process. It became a monstrous tug of war. I was convinced the boy was dying.

It is almost impossible to describe such a moment. No words exist capable of translating such terrifying emotions. But often a tiny little speck of time gives heroes their opportunity to demonstrate their courage.

Ralph threw the first thing he could grab at the cat, whose chest was pressed against the boy's back, and ran at him with a short, accelerated rush. The intent animal didn't even twitch as the brick hit him, but the sudden advance by Ralph startled the tiger holding the boy's leg, and he let go. Then in a fit of sheer panic he spun from a low, crouched position and sprang wildly into the air.

Ada now arrived, screaming jibberish and issuing orders to everyone in Polish. No one could understand her, and she couldn't get it together enough to switch over to English. She sounded like Woody Woodpecker on helium.

Out front there was no pandemonium, no hysteria or uncontrollable reactions. The show continued as though nothing out of the ordinary were happening.

The frightened airborne tiger flew by with aerial grace, landing on top of Sampson, who reacted much like a cow dislodging an irritating horsefly. The elephant threw around his mighty head, shook his massive body, and with a bundle of controlled power caught the startled tiger (who, I suspect, had mistaken Sam for a mountain) by the underbelly with his tusk, then, with a forward thrust, tossed him into the air from whence he had come.

Ada, not understanding why no one was responding to her foreign tongue, opened wide the nearby closed cage door, and as the cat landed on all four feet directly in front of her, she stood with a shovel in one hand and a crop in the other, calling in a voice that would put a fog horn to shame for the tiger to enter his house. With little insistence, for I don't think the confused cat was ever so glad to see anyone ("Listen, Ada, it was all Raja's idea. He made me do it"), he ran with relief for the open door and was locked safely away. One down, one to go.

The other predator had managed to drag his morning meal over to the side of the same 10'-x-10' cage. Blood now gushed from the boy's legs; his young face was deathly pale as he hung from the clenched, crushing jaws. Again in a small agonizing whisper he called "Help me! Please someone help me!"

Along with Ralph, who was physically trying to pull the tiger off, and Fess Reynolds, who was whaling away with an aluminum shovel, a keeper began beating on the cat with a flimsy bamboo-fan rake.

Then the animal suddenly released the gore-stained youth from his death hold, and the boy sat motionless, locked in a frozen position. The impassioned Bengal barked and flew in rage at each of us, holding off all opposition in lunging, threatening insanity. I hit the surprised tiger on the head with my clipboard. Can you believe it? Then I yelled something ridiculous like "Don't call me any of your family names!" Fortunately I was still young enough to believe I'd never die. So I didn't faint and embarrass Ralph, who was yelling, "Don't move, Toni. Don't move!" Move? I'd probably never move again. I was so rigid you could have hit me on the head with a baseball bat and I would have shattered in a thousand pieces. The aggressive, deranged animal turned unexpectedly and brushed by me as he ran out the back of the theater, around the corner, and into the public restroom. The men and Ada were in hot pursuit. The boy now stood and tottered toward me like a dead person rising from his grave. It was such a frightening sight I thought fleetingly about hitting *him* on the head with my clipboard. (Why do I say things like that?) His bloody arms, shredded, stretched out far beyond sadness and despair. And having gone beyond a certain threshold of pain, he fell into my arms.

Why there was no lineup at the ladies' room shall remain a mystery to all of us. Usually the place was jammed with mothers and their cross-legged kids. I can only surmise because the show was still in progress, they had remained in the theater. Here among the tidy bowls and toilet paper the enraged cat was caught, but not before he had tried to kill the other tiger in the mirror.

Thank heaven the boy wasn't as damaged as he looked. He was nonetheless punctured severely in several places. Although he remained bedridden for several weeks, it didn't take long for him to heal.

11
That's Three, Tinker

There is, nevertheless, a certain respect and a general duty of humanity that ties us not only to beasts that have life and sense, but even to trees and plants.

–Michael De Montaigne

Tinker Johnson looked at me with his usual bland expression. "Well, Missie, this time I got me sompin' yer man's gonna love. Just lookie here." And he gestured me toward a large box. Having a preoccupation with curiosity, I followed closely behind, lecturing Tinker all the while: "I have no authorization to buy anything, Tinker, and personally, to tell you the truth, I wish you'd stop coming here. It seems every time I see you I get into some kind of trouble." With his habitual expression, he turned around to face me, and, never flicking an eye, he opened the plywood door. Before I could protest, Tinker had by the tail and immediately upon the ground an enormous monitor lizard. I jumped back several feet, expecting the thing to attack. But the huge reptile barely moved.

"Okay, Tinker, what's wrong with him?"

Tinker reached down and placed a little collar and leash on the now lively beast and, patting his side, coaxed him into walking, which he did in a wriggling, angular pace. "Here, Missie, you hold this." And he handed me the leash. From where I stood on top of his flatbed truck, I didn't imagine the danger to be great; so I bent low and held the vertical strap. "Missie, there ain't nuttin' fer ya to worry about; old Clyde here ain't gonna hurt ya none. Why, he's as gentle as them other critters of yers. I only stopped by to show him to ya 'cause I knew how much that man a yers likes reptiles. Clyde here's one in a million, ain't ya fella?" The dirt was being crushed under the bulk of a thrashing tail as he spoke.

"Gosh, Tinker, he hasn't opened his mouth. Come on, what's the matter with him?"

"I swear there ain't a ting wrong a him. He's just got a good nature, and it seemed fittin' that man a yers should have him. Come on, Clyde, let's show her how good ya are."

The six-foot lizard whipped along with the old buzzard by his side, and although I had a brief moment of apprehension, I couldn't help thinking that while I was saving the lizard from the likes of Tinker Johnson, it would also be a wonderful surprise for Ralph to come home to. I began to capitulate as Tinker reassured me, though I knew the whole time in my heart that I was being taken in by a beguiling expert. I could not help admiring his act. And after he had finished the ethics-ridden speech as to why Clyde and I should be united, I must confess he had me convinced. Tinker invited me to take Clyde for a walk and see for myself what a fine fellow he really was.

The lizard and I twisted along, experiencing only minor points of contention. I found the unusual circumstance of taking a lizard for a walk rather special, and the more I studied the reptile, the less he looked like an alien and the more he resembled someone I'd like to know. Besides, he was so incredibly good. Clyde was a must! Ralph might be upset at first, but I was sure as soon as he came to know Clyde, he would be forever beholden to me.

Yes, Clyde needed us. I did not require confirmation from the flesh forest of trainers. One could see he was a congenial companion. Clyde and I swiveled up to Tinker. All the while I tried desperately to hide my excitement, for I knew I had really come across a "find." If I could only keep my emotions under control, maybe I could wheel and deal with Tinker in the same way in which he always chisled me.

"Tinker, remember the hyena you sold me? The one you said came from a little old lady in Washington who kept him for years as the family pet, and only when her husband died and she had to sell her farm did she agree to part with the dog? And you said there wasn't anyone in the world she would rather see him go to than me? And you kept telling me how sweet he was and how my 'way' with hyenas would transfer him over in no time? Remember, Tinker? And the Pasteur treatment? Now that one I know you can't forget, since pain was involved. I'm sure the memory is etched on your brain. Well, the 'sweet little dog' gnawed me nearly to death, and the shots, Tinker–need I go into detail regarding the shots? No, I think not. So you see, between the doctor bills and the psychologically devastating ordeal, as I see it, you owe me! I would accept Clyde as payment in full, and we'll call it even. How about it? Is that a deal?"

"Well, Missie, you drive a hard bargain." He stroked his unshaven face, shook his granite head and said, "If'n ya give me twenty-five dollars, realizing of course that ol' Clyde there is priceless, well, I'll go fer it cause yer old man's done me some good turns."

"Done, Tinker!" And I pulled forth twenty-five new ones and sloshed through some more trivial anecdotes before he drove away.

Clyde's tail agitated as I picked up his thirty pounds and placed them in the back of my wagon. I took him to the house and turned him loose in the hall, but not before tying a huge red bow around his leathery neck. He looked absolutely ridiculous. I could hardly wait for Ralph to come home, though I was acutely aware of the potential danger the animal could cause. Reptiles like Clyde were known for their nasty temperament and aggressive tendencies. Clyde repeatedly struck the floor with a stroking beat, occasionally thrashing about and jerking his head, flogging the air, yet never once showing his fierce teeth, beautifully mannered. It was an impressive sight.

"Just look at him, Ralph, just take a look. Have you ever seen such a lizard?" Clyde whipped around with keen interest and mildly observed me as I snapped his leash and collar on. "He's leash broken!" I revealed with glowing delight as Clyde and I slithered onto the kitchen floor. "Not once has he offered to bite me, not once. Don't you think that's amazing? I mean you know how violent they can be. You're always telling me stories about *Journey to the Center of the Earth* and the *Incredible Shrinking Man* and all those dinosaur movies and how difficult they were to do."

Ralph did not look thrilled. In fact, I would go so far as to say he looked increasingly annoyed. With my bottom lip protruding, I asked, "Don't you like Clyde? Aren't you pleased with his sparkling personality and mild, sophisticated manner?" With all the grace of a Southern gentleman of the pre-Civil War era, Ralph advanced upon me, kissed me upon the nose, stroked my hair, and with a wide-eyed gaze, looked into my eyes. Taking a great, heaving breath he asked, "Do you have any idea why Clyde is not trying to eat you alive?"

"He's a very gentle creature; it's not in his nature." And I reached down and stroked the bug-eyed fellow.

"No, Darling, no, that is not why Clyde is so passive."

"I don't know what you're talking about. He and I have been together for nearly seven hours now, and he's been just as pleasant as can be. You're just mad because I got him from Tinker Johnson."

"No, Toni, that's not the only reason I'm upset."

"Then what's wrong? I don't understand."

"I'm upset," he confided with an ingratiating smile, "*because Clyde's lips are sewn together*!"

That's three.

And if I ever see you again, Tinker Johnson, heaven help us both.

"Oh, how gross. Poor Clyde! Do you think he's in any pain? Ralph, you've got to hurry and get that stiching out! Oh, I just can't believe anyone could do such a monstrous thing!" I bent low and gave Clyde a friendly, empathetic tap on his bulging sides as he arched his V-shaped head and thumped his elongated tail. Ralph began to unravel the sutures, his indignation increasing over the pathetic condition of the lizard.

The sight of Ralph slowly pulling the stitches out of the animal's mouth as I held him by the snout made me feel utter contempt for Tinker Johnson. Carefully he worked the thin hemp through the pin holes, trying to lessen my anxiety. "As grotesque as this appears and as inexcusable as it is, there is one consolation for Clyde. He's a cold-blooded animal, and his pain threshold is considerably higher than ours."

"But, Ralph, I can't understand why anyone would do such a terrible thing."

Clyde threw me a calm-eyed glance as one more thread came loose.

"When they ship these fellows into the states, they crate a lot of them in the same box. To keep the lizards from fighting so that they will arrive in one piece—also to keep them from biting—the trapper sometimes lowers the body temperature of the animal and sews his lips together in this ruthless fashion. Remember, the people who make a career of catching reptiles are only interested in economics. Anyone involved with animals is unfortunately identified with this kind of cruelty. The struggle to stop such barbarism reaches an impasse when people afraid of reptiles don't care enough to put a stop to it. But what an infringement it is on an animal's rights."

Clyde was now working up a lather and blowing bubbles from the side of his green mouth.

"So we start in our little way to eradicate misconception, accepting our commitment, trying very hard to follow through, but it takes enormous interest by citizens to bring this kind of atrocity to an end. And one of my vested interests is to protect these guys from savagery like this. But the cultural legacy continues to work to our disadvantage, because there are more people who don't care than care.

"There, the last one's out. Don't let go yet; I'll have to put some hydrogen peroxide in the punctures to prevent infection."

He returned from the medicine box and poured the cold antiseptic solution over the little wounds; Clyde began twisting and turning, straining against the bridle of my hold. I hung on desperately, pinning his tail down with my knee and struggling to keep the squirming creature from getting free.

"Ralph, I don't think I can hold him any longer. He's been so good; couldn't we let him loose now?" I grunted with growing muscular contractions.

"I don't think you will find Clyde particularly grateful," he said with concentrated medical concern, dabbing cotton on the expanding mouth.

I was amazed at the sudden burst of struggling strength as, with a ripple of his tail, Clyde pulled away from me, and before Ralph had an opportunity to grab him, he had spun completely around. I collapsed, folding to the floor, where I ceased to function. To my horror, I still held the lizard by his snout, which was now directly on the same level as *mine*! I had him only by my thumbs and forefingers.

Forcibly he unfastened my slippery hold, and from that moment all I saw was a red, gaping canyon of dedicated ferocity, rimmed with a gleaming edge of razor-sharp teeth. The huge moist mouth advanced threateningly, erupting forth a horrible "hisssss" before hurtling itself at my face. But thanks be for men who prefer their wives without masks. Ralph had him by the tail, and Clyde, just one bite away, was suddenly airborn and flying around the room, whirling around and around like a green string on the wing.

In my hypnotic state I watched with fascination as Ralph and the lizard spun together. "Toni, don't just lie there; get me something for him!"

I vaulted to my feet, leaning back as the mouth flew by. "Where's the needle and thread?"

"A pillow case. A pillow case! Will you *please* get me a pillow case!"

We eventually got old Clyde put away. And through many scenes of high comedy and dizzying fatigue, Clyde eventually became an established member of the reptile order of Africa USA. But he was never quite so good as when his lips had been sewn together.

12
Blood, Sweat, and Steel

And so geographers on African maps
With savage pictures fill the gaps,
And in all uninhabitable bounds
Placed elephants instead of towns.

–Jonathan Swift

We sat at our table having scrambled ostrich eggs for breakfast. The early morning sun was pale, whitewashed by a low-hanging mist. Animal sounds drifted through the open windows, clear, vocal, and strong. We captured for a brief while an existence with diminished anxiety and other social disorders.

The walk to the ranch down our country road through heavy, evaporating air was enough to lighten one's head; it gave to us a kind of impractical fever, rich in imagination. It seemed to me that my days, even at their very worst, were a great blessing. A certain character had begun to form on our faces. There was from all outward appearance the supportive evidence that I was deliriously happy, practically all the time. But that day it seemed that nature, not totally content with such a creature, began to place certain burdens against my condition, and for a brief while I began to wonder if my life had suddenly turned into some kind of endurance test.

At 8:30 a.m. I assembled the students on the lawn outside the nursery, under the trees where the overpowering smell of summer—of new grass and warm earth—enveloped us. The birthing of a glorious, fragrant season was upon us, and the students were bursting with eagerness and youthful vitality. As usual I was bombarded with enthusiastic questions: "Can we bathe the lions?" "Can we have a swim with the elephants?" "Can we wrestle the bear?" "Can we ride the giraffe?" "Can we . . . ? Can we . . . ?"

"Hold it! We can not! At least not for a few hours yet. First, you're going to assemble your own cat leashes; then you're going to learn how to splice rope and make your own halters and leads. All this is to be followed by a fabulous knot-tying demonstration, in which I will attempt a perfectly vulgar display of my great skill. After which will come play time, boys."

Being an instructor at the ripe old age of twenty-six to a class of twenty-five aspiring men of twenty-one years and upward was, as one might suspect, a bit of an anxiety. There were times when my only defense against them was my supreme dictatorship. When their questions and critical analyses bisected my brain, it was inspiring to have the power to call a halt to the drilling inquisition and shout orders in my military fashion (a language they all understood): "All right, men, that mountain needs a good cleaning!" Commands of that sort were always enough to change the focus of conversation, and let me tell you, we had the cleanest mountains around!

Under a hush of concentration and frustration, knots were being tied every way but right, and so that I would not injure delicate egos, I tried as well as I could by "show and tell" to correct their mistakes.

"The end of the rope is that part in which the knots are tied. The remainder of the rope is referred to as the 'standing part.' When our rope is new, we must 'work' the stiffness out. We do this by pulling, stretching, or twisting it to render it more supple. Like so. Here are several knots we use in our business more than the others. They are the square knot, the bowline, and the two half hitches."

As I began again by tying the two half hitches around the stick, the boys mimicked my every move. The intense focus of concentration was suddenly shattered by a violent, incredible blast. It shook the spot where we sat with such unexpected force that we sprang to our feet as if by ejection. The earth erupted with a dreadful vibrating disturbance, in a sound so final that one suspected the sky was falling down. Recognizable among the foreign noises was the crunching of steel and the whine of metal as it twisted and tore. Pieces of rock flew like lethal insects through the confused air above us, and a mushroom of black dust rose from in front of the *Daktari* set. Someone screamed, "It's a plane! A plane must have crashed!" As we ran in shocked silence toward the cloud of dust, the thought took hold of me like icy hands upon my heart, "It's gone down on the animals! Please, God, no!"

Cast and crew from the set were standing stunned, waiting for the cloud to lift, waiting to see what had happened. My heart pounded like the drum of a fast-sailing slave ship as I walked hesitantly toward where the Beverly Hills section had stood, housing over twenty celebrated animals. I couldn't see a thing through the unhealthy black mist. Steam and oil sprayed all around us, and so thick was the smoke screen it was impossible to distinguish what the scattered piles of debris were. I was afraid to breathe, to move, to think; and just as Ralph came to a skidding halt before the frightening scene, the cloud rose and revealed through a sooty veil a capsized steam-powered crane car. It lay wedged against the embankment, barely supported by the half-crushed dwellings beneath its 100 tons of iron. I felt nauseated as we rushed toward the wreckage. The terrifying hush had ended; screaming and shouting and crying began.

I looked as though I had come dressed for the occasion; from my neck dangled five cat chains and several ropes. In the split second of assembling his thoughts, Ralph grabbed one of the ropes, nearly hanging me. "Toni, give me one of those chains! Somebody get the bolt cutters—and hurry!"

The iron horse lay puffing and spouting as though the great man-made machine were taking its last breath. Under its broken weight was a depression of unknown horrors. We began forcing open the mashed enclosures, cutting away the six-gauge chain link with enormous cutters. Everything was smeared and soiled with oil. The odor was stifling, and as we tore at the holes in Serang's cage, one very black, greasy tiger, filled with overwhelming enthusiasm for having been released, slid on top of Ralph, licking his face and slobbering all over his head. Ralph clipped the leash around Serang's neck, and Serang clung by his side like a shaking mass of slippery glue, looking up with quizzical, jittering eyeballs, as if to say, "What the hell was that? Did you see it? Did you see it? Huh? Huh?"

I gave him a great, petroleum hug as some unrecognizable face in the crowd took him away, talking to him in baby talk and giving him reassuring pats on the side.

In despair we moved on to the next cage, where a rapidly panting, tail-thumping wolf gave an overconfident yawn, then searched the disturbance above her with vague resentment. Her door opened with unexpected ease, considering only three feet of space remained inside. Sabra wiggled all the way over, then sat composed, reining in her restless energy. In her optimism I found my calm.

The Iron Horse, puffing and spouting its last.

Down the row we went, releasing the small group of brave souls, expecting the worst but finding in the visual gloom that other than a few minor scratches, the only injured animal was Sampson the lion who had a slice in his front right pad. It seemed incredible that from this unbelievable wreckage the animals had been salvaged. For the most part they were remarkably under control in the midst of such pandemonium. In my maudlin sentimentality I believed they, like little children who had implicit faith in their parents, knew as long as we were there everything was all right. In a way I celebrated this accident for having been kind.

It wasn't long before the arena had been evacuated, and while I organized baths for the bewildered, oily lot, Ralph climbed to the railroad tracks and found a man's blood-stained body outstretched, limp and dying. He never regained consciousness. We learned he had been a man of splendid courage. The investigation proved he had ridden the runaway car for miles, trying to stop it with a hand brake. He knew the curve in the rails was at our place; so he stayed aboard (even though two other men had jumped off miles back), doing all he could to avoid the inevitable event and so fell victim to the Iron Mammoth's destiny. Like the captain of a sinking ship, he had chosen to be with it at the end and had died with the powerful steam car he had come to regard as his own. I understand that kind of love for machines, and although I may never risk my life for one, I can see why some men do. I know many men whose happiest hours are spent covered with oil, surrounded by locomotion and with steam piping through their fervent veins, as much a part of the magnificent scene as are the great machines. And, like my dear friend David Shepherd—the world's greatest wildlife artist and well-known collector of steam engines—I can empathize with their affection for such giants, for I love elephants.

13
The Lion in Winter

The dry wind scatters grains of sand
As egrets sail 'round like falling snow.
Down on proud Egyptian herds they land;
Among Solomon's mighty herds they go.

–T. H.

On a stormy electrical night nature lashed out in ferocity as bolts of lightning flashed all about us and sinister thunder roared. Sleet pelted the windshield of our car with such force it seemed as though it might pop.

In front of our house had stood, for who knows how many such winters, an oak tree, dear to us through long acquaintance. Those strong branches had rocked our Tana in a swinging tire as she grew. They had held the reins of King and had provided for us a canopy of soothing shade. A choir of birds had lived among its foliage. Under her mistletoe I had been kissed. She had given pleasure to the eye under the glaring sun.

Now, as we slid onto our driveway and rolled toward the house in the thick slush, we saw that the roots of the old tree that had been buried so deep were now exposed, darkened by decades of soil and lying just at the entrance to Zamba's room. Like a great old colossus the tree, split by lightning, lay in two sections, its greater half across Zamba's room.

Ralph and Zamba playing. (Photo by Richard Hewett.)

"Zamba!" Ralph cried out with a sound that made my heart stop.

We burst through the front door, not knowing what to expect. Zamba had gathered himself into a far corner, trying to make himself as small as possible, the tree and portions of the roof scattered about him.

With a sigh of relief, we called to him, and he bounded with huge leaps across the primeval room. He flopped over into Ralph's lap, scolding us for leaving him alone and making us promise it would never, ever happen again. Having sufficiently bawled us out and feeling a little better about things, he relaxed and began to lick Ralph's arm with his abrasive tongue, and so changed the too-much talked-about subject.

Between swats from the huge front paw, Ralph checked the amber eye. Now that Zam had released his frustrations, he wanted to play. A strong yet gentle forearm wrapped itself around Ralph's neck and pulled his head onto the curly underbelly of the burly giant, and the black tufted tail swung around and walloped him on the back.

"Come on you moose, let me up!" And for a while they wrestled as those kinds of animals do.

"Okay, that's enough," RDH issued.

Zamba moaned that he couldn't have *any* fun, sat up, placed a somber chin on his paw, and began to pout. Ralph comforted his spoiled child with some reassuring pats, then pulled at the cat's sleek coat. A handful of golden hair came loose. Ralph stared at it.

"What's wrong?"

"I'm—not sure. I just have a funny feeling. Oh, it's probably nothing." He reassured me and shook his head. "Just the excitement from the storm. But tomorrow Zam's going to Marty for a checkup. Come on, you big old lazy thing. The drama's over. It's time for bed." The aged citizen of the world slept comfortably, draped across us that night.

The next morning we had to shake Zam awake. When he finally stood and stretched, he stumbled; and when he walked, he walked a crooked line. We took samples of everything—urine, stool, saliva, blood, all of which were raced to the lab, labeled "Urgent reply requested." Zam spent the day in the office with a very anxious Ralph. In a matter of hours the report came in by phone. We were told by an uncompassionate voice that he was suffering from severe uremic poisoning.

We were dumbfounded. Why, the very thought of anything ever happening to Zamba was inconceivable. We spent all that day and into the night with doctors in consultation. We tried everything, but old Zam declined steadily. Ralph stayed awake with Zamba that night, and when I could no longer keep my eyes open, I stole an hour's nap.

The sunny rays of dawn shot through the lion's room and warmed me awake. As I opened my eyes, to my dread Ralph sat staring at the wall, holding the lion's great head in his lap. His arms were wrapped around the cat, and he rocked him like a baby. Tears streamed down his cheeks. I started to speak, and he buried his face in the black mane of his best male friend, sobbing uncontrollable surges. Then he closed forever those amber eyes.

We buried Zamba under the ancient oak at Beaver Dam. As we lay him on the soft bed of straw, Ralph placed his rawhide neckpiece with him. It began to rain as we covered the grave, and the skies cried with us. We went home and washed his oil from our hands for the last time.

When Zamba died, the lion stayed with Ralph, but a lot of the man lay under the tree, and everything became reminiscent of Zamba's time for a long while afterward. It rained a lot that year, and when the tree Zam lay under was struck by lightning and split in two just as the one on the eve of his death had done, it seemed a fitting marker.

And now all the oaks had bodies of our friends buried beneath them. When Ralph and I took walks through our land, we would say Zamba lies there, or Scalla lies here, or Toby and Jenny lie on the hill; the park soon became a graveyard, and we were afraid to step on the sacred ground lest we stand on someone we knew.

14
A Dissolution

Their cause I plead—plead it in heart and mind.
A fellow-feeling makes one wondrous kind.

—David Garrick

In 1968 our partnership with Ivan Tors came to an end.

We parted from him as friends, but the new group were doing their best to obtain what had always belonged to us—our remaining half of Africa USA.

It became for some of the employees a time of choosing sides. Everyone wanted to be on the winning team, for many of them felt their lives were being uprooted by the Helfer-Tors breakup. I suppose we naively assumed from those who had been close to us a kind of allegiance, but the unfulfilled desires of some of them were being stoked by a neurotic society that promised rewards for selling out. Those negotiations certainly separated the strong from the weak.

After a long course of reasoning we succeeded in retaining the property and the bulk of the animals; full ownership reverted to Ralph and me. Those litigations had been emotionally devastating. We were mentally exhausted from the difficult bargaining. The terms and conditions of the

Toni and Cyrano, the anteater.

final agreement had given us strong subjective feelings, for many of those fabulous animals, whom we had all profited by, were penalized by the dissolution, winding up paying the piper. I would rather they had died than have been shipped off to Florida with that group of men, for these new people knew nothing of Clarence and Judy or Gentle Ben and my two red-haired orangutans, Hannibal and Genghis Khan.

But they could, by exploitation, make a good deal of money off them.

Toni and Ralph with three-year-old Dandy Lion.

It's very hard not to cry when I think of the sad fate held for each of them, and I call to mind a magnificent verse by Karen Blixen from her book *Out of Africa.* She stood on the Mombasa shores and watched as two giraffes, loaded in wooden crates, were waiting to be shipped from their homeland to a port far across the sea.

"Good-bye, good-bye. I wish for you that you may die on the journey, both of you, so that not one of the little noble heads, that are now raised,

Cobra (a Harpy Eagle) and Ralph in training for the movie Harpy.

surprised over the edge of the case, against the blue sky of Mombasa, shall be left to turn from one side to the other, all alone, in Hamburg, where no one knows of Africa.

"As for us, we shall have to find someone badly transgressing against us, before we can in decency ask the giraffes to forgive us our transgressions against them."

For a long while we had spoken of opening Africa USA to the public. Only recently a major studio had approached us with a tempting proposition. If we joined forces and became an extension of their own successful operation, they would bus their visiting tour groups out to our wild-animal ranch. Not only would they invest a considerable amount of capital in the business, they would ensure our success by the sheer numbers from their tour business. We considered their offer with growing interest and began with renewed vitality to prepare the Saugus compound.

Toward the end of 1968 we bought a beautiful home in Sand Canyon, ten miles down the road toward Saugus. We had completed four prosperous years, and having worked so hard for so long, we decided to reward ourselves. The house was such a contrast to the little cabin that it took me a long while to adjust to the fact that I was not simply a visitor here. I wandered through the sixteen rooms for days and curled up in every one of them to get the feel of the place. Through the years I had collected enough artifacts from Africa to start a continent of my own. I began decorating our home with fragments from a primitive time, and it wasn't long before it resembled Kenya as seen through memorabilia. Around the swimming pool I stood enormous ceramic statues of lions and tigers and leopards, succeeding in scaring the wits out of the pool service when they came to clean.

The holidays in our new house were celebrated like the Queen's Silver Jubilee. From Thanksgiving on was a month of joyous ceremony. And of course this Christmas held even more magic than those in the past; for we were now just thirty days away from opening the ranch to the public, and for once we had enough to money to invest in our goal. So we spent it in a special sort of way. Rather like ornamenting a live Christmas tree, we put the final touches to the decorative landscape and with signs added a history that told the tales of the land now known as Africa USA. Intoxicated by our expectations, we walked with increasing pride through the park, as a passionate collector of great art might view the personal workshop of a Cezanne, a Renoire, or a Van Gogh. All Ralph had worked so hard to achieve in his life was at last about to come true; so Christmas this year meant a good deal to us all. It was a sacred gift—a future filled with promise.

Clad in the spirit of such a season, we took many of the animals into the wards of children's hospitals. The chimps dressed as elves passing out presents brought squeals of giggling delight from the little people who resided there. A nurse accompanied us from bed to bed, summing up their young lives. "At the most he has six weeks to live." "She'll never be able to walk again." "He lost his legs in a car accident." It seemed to me her daily pessimism was in itself an incurable disease. The children had more control over defeat than those of us who were suffering for them. Their courage elicited in me a secret sorrow, and my problems seemed very small in comparison with theirs.

That winter at Children's Hospital a little boy named Ricky Warren left a lasting impression and a kind of spiritual awareness with me. He sat in his wheel chair, thin and ghostly pale, almost translucent, with barely a sign of life left in him. His gaunt head rested to the side, and his arms hung limp before him. Ricky was the kind of child that made you choke when you looked at him, and you prayed you wouldn't cry when his turn came to meet the animals.

As we stood before this grey boy, I felt very tall and at the same time very small. The chimps treated him as one of their own and seemed by their presence to breathe some life into the child. Mac pushed his chair around while M'toto found increasing interest in his skeletal hands, first lifting one then the other, and stroking the withering, dehydrated skin. I made a motion to stop the chimps from bothering him, but he said, "Please, let them be." It was then M'toto climbed on the foot rest of the chair and looked with a penetrating stare into the boy's sunken eyes. It gave me goose bumps, for it was like witnessing one brain entering another. When the chimp was through searching, he climbed into that withering lap and hugged Ricky Warren like one whose cherished stuffed toy had come suddenly alive to comfort his friend. That chimp knew; he knew.

Ricky smiled, then raised those heavy arms and hugged M'toto back. I thought such a grip would break his frail arms. He turned to me, his face warm and shining, still holding tightly to M'toto, and said, "I've always dreamed of knowing someone like him."

Ricky and M'toto corresponded as pen pals for the next few months. M'toto and I put much into the answering of his letters, in this way perhaps hoping to prolong Ricky's life; for at the end of each letter we requested an answer, giving him one more reason to live. Through the years I have given much thought to children like Ricky Warren and to animals like M'toto who bring such happy moments to them in their diminishing hours. The capacity of some animals to show empathy and to give a dying boy joy is a very special medicine, a unique healing power all its own.

The senior citizens' home was a place I found lacking a certain humanity. These lovely old people had in many instances been committed here by children who couldn't find the time to care for them. They found, even more than family abandonment, the deepest pain of all being separated from their pets. The homes would not allow their cats or dogs or birds to come with them. What kind of a society is it that cannot see the great need for old people to keep their animals? Those animals are probably the last link they have with touch and affection. Are we so caught up in our own youthful ignorance that we fail to see human qualities? Thank you very much, but I would rather someone put me mercifully to sleep than strip me of those I love, who in turn love me. And that is exactly what most of the aged clan do, choose either to die themselves or put to sleep their faithful pets. Someone should write bylaws for the "Society for the Prevention of Cruelty to People."

To be with the blind when they touched their first lion was a rare and intimate moment of sharing. And the orphans—how they loved the animals paying them a visit at Christmas. Many of them said it was the best time they had ever had. I had a sense of loneliness when I was there—of an existence in a community of little people whose very lives depended on faith. They did not take their exile lightly. I thought many times of adopting, but with all my animals, not enough of me was left to go very far. And Ralph, who has always had a great love for children, would, if finances had allowed, taken on the whole orphanage. I felt a sense of loss when I left, as if I'd left something important behind.

There were dozens of benefits—for this charity and that charity. The animals were always an inspiration at fund-raising events. And the parades—how many parades Ralph and I rode in, sometimes as Grand Marshal with Clarence and Judy riding in the seat next to us in our open-top convertible. But more often than not we were transported by elephant power. One year during the organized confusion of the Santa Claus Lane Parade, Hollywood Boulevard swarmed with a wide swath of excitement. As the bands fell into procession, they played yuletide carols, and the crowds sang "Jingle Bells" on the sidelines. In such an atmosphere you couldn't keep from smiling and wishing the world would stay like this forever. You were friends with absolutely everyone, and no one person or group was dominant over another. Religions were as one; of Christian or Jew no one thought the better. It was a special night of good will toward all, when we were dependent one on the other for that brief spell.

Bill Burud was hosting the spectacular parade for channel eleven. We had supplied animals for his shows through the years and were on friendly terms. Tana sat in front of Ralph; I was behind him, and Rosie the elephant was under us as we ambled along, waving and tossing candy canes to the cheering spectators. The serpentine procession came to a standstill as Bill and his microphone stepped out to speak with us.

"And here, coming up on my right, ladies and gentlemen, is the man responsible for all those great animal shows you see on the screen–the creator of Affection Training–Ralph Helfer and his lovely family."

After sixty seconds of animal gossip, we and the parade marched on. It was a strange view of the world, riding at such a height. From where we sat I could see the dark-haired kid with a sling shot perched on the window ledge, and I was suddenly stricken with apprehension. He pelted Rosie in the hind end as we passed by. Rosie began to shake from side to side like a building on its way down in a great quake. Try riding a bucking elephant some time. It makes the rodeo look like kidstuff. As she began to trumpet and scream, I told Ralph what had happened, and at the same time I heard a lady cry out, "Oh look, isn't that cute? The elephant wants to play." I don't think so lady! Poor Rosie reacted as if she'd just been attacked by a bee brigade. Our mouths now frozen in smiles, we tried to maintain a sense of calm as we hung on for dear life; for as Rosie began to run, Ralph had his hands full trying to steer her in the direction of a side street. We whistled toward the unsuspecting crowd blocking our path. It parted with startling speed. I called out (still smiling as though this were part of the show), "Beep–Beep!"

Most of the people were applauding and laughing as we lumbered past. One concerned citizen yelled, "Are you all right?" to which I replied, "Which way to the nearest pachyderm powder room?" They thought that was terribly funny, and I'm sure they laughed for days about the elephant who had to go potty.

It was lucky for one and all that the streets on the other side of Hollywood Boulevard were empty that night. Tana was having a wonderful time on the runaway beast. I resented that enormously. A derelict looked rather alarmed as we breezed by.

The saddle began to slip and slide, and I had a strong desire to abandon ship, but Ralph jumped off with Tana first. Me first. Me! Me! Me! How gallant! I was riding side saddle–all the way over on Rosie's side–my head spinning as the streets and buildings blurred by. Ralph was running like Bruce Jenner, trying to stop Rosie, who was putting her all into streamlined elephant power. The four-ton hulk was leaving footprints in the asphalt the likes of which Groman's Chinese Theatre had never seen.

"What about me?" I screamed.

"Hang on, Toni," my hero responded.

Hang on? Do you now know where I was riding? Practically beneath her belly! The saddle had fallen nearly all the way down, and my legs were locked around the metal hand railing; I looked like the awkward lady on the flying trapeze. Looming above me was a living, quivering mass. If for some reason she should decide to sit, it was "squash time"; she would come down on me with a sickening splat. It was a ridiculous looking sight. And besides being terrifying, the position was somewhat painful. Why were these things always happening to me?

"Don't panic, Honey; she's just scared. She can't keep up this pace for long."

"Can you?" I interrupted. I hope, I hope, I hope.

I was hanging on so tightly that when the time came, they would have to pry me loose with a crowbar; I was welded to the saddle.

I could see from my nearly upside-down position a gang of young boys huddled together on the sidewalk snorting marijuana. (Is that how it's done?) They looked at us strangely as we barreled by.

Rosie came to a sudden stop, unfortunately not before we had demolished a white picket fence and several flower beds. I dismounted with my usual savoir faire, by falling to the ground from the seventy-two-hands-high figure, and scooted rapidly far from her feet.

We walked, soothing Rosie's neglected feelings for nearly a mile, back to the semi that had been parked at the beginning of the parade. Other people were walking their dogs.

15
The Flood

Hopes, wishes, aspirations, ponderings
After ages and ages of incrustations
Then only may these songs reach
Fructation.

–Walt Whitman

The rains began a little early that year; they were falling with a steady increase in late December, never letting up for very long. Bad weather has a way of bringing people closer together; nearly every night the group from the ranch drove over to the the house, and we popped corn and toasted marshmallows around the living room fire, creating an atmosphere of a tightly knit family. Everyone had been so used to our living across from the compound that our departure at the end of every working day came as an unexpected void in their lives. For a while, like little lost sheep, many of them followed us down the canyon; and so that we would not cut them off suddenly, we'd let them spend the night in sleeping bags and begin their days in our new home.

On a January night in 1969 thunder rattled the glass of every window in our house. When the lightning flashed at 3:00 a.m., our rooms were as bright as if it were day. With powerful illumination the spectacular storm pounded down.

We had spent $75,000 completing a flood channel in 1967. It had been approved by the county, and all powers in control believed that if the so-called 100-year flood were to hit, everything in the canyon but Africa USA would be affected by it. So we felt secure in the knowledge that we had moated and protected our land. Every winter the stream rose, forgetting its bounds and spreading perhaps twenty feet farther than it should. We drained the lake at Beaver Dam, leaving enough bottom water so that the fish would survive there, and we rerouted the little stream to the wide channel below. With this advanced planning we prepared for all storms, but *not* a much-talked about but never realized flooding.

At 6:00 a.m. the phone began shrilling. It was Charlie, the night watchman. "Hurry and get over here! The river's rising! And they say the dam at Little Rock is going to be released on us before it bursts!"

"Charlie, stay on that phone. Get everyone you can on their way. we'll be there in ten minutes," Ralph directed.

Struck with a sense of urgency we grabbed blankets and clothing, boots and lots of socks, threw some food into a sack, grabbed a sleeping bag on the way out, piled into the car, and peeled out of the driveway into the raging storm. And so began a mute ten-minute drive. As we crossed the bridge over Sand Canyon, we could see the little stream had become a swollen river, elevating and spreading under the dreadful inundation.

We spun on to Soledad Canyon Road and two miles further began to turn the hairpin curves, swerving here and there to miss the fallen boulders as the river violently washed along the canyon beside us. Past River's End the road was collapsing, eaten away in parts by the deluge. The rain now came down with such blinding force that, as we bounced over the eroding asphalt, the road was invisible. For all we knew there was no road ahead.

Through the tunnel, around the bend by White Rock Park, the highway was almost completely gone. In only a matter of seconds Soledad Canyon would be locked in on that side. With a deafening roar the road we had just driven across caved in and became part of the torrent below. As we sped past the Red Fence and Bert's house, we could see that Oasis Park, which sat in the middle of the channel, was in for terrible trouble. The Wilsons' farm and the detention camp were on the right side of the road. In front of our old house we came to a skidding stop, parked the car, grabbed things we had brought, and started across what was left of the entrance to the park. The dirt above the enormous culverts was mostly washed away. The great cement tubes were beginning to turn and give as trees and debris began to pile against the only way in and out of Africa USA. Gripping each other by the hand we made our way, hopping from drain pipe to drain pipe, followed by a few employees who had come in from the Acton side of the Canyon. As the timbers accumulated to the side of the man-made bridge, the waters rose and backed up across the 150-foot aqueduct. The once-in-100-years flood was on its way. I suddenly had a longing for places I'd never been, and for an ark, above all else an ark. When we reached the compound, some of the employees stood by, immobile, waiting for the commands of leadership.

"Isn't this awful–where shall we start? What shall we do?"

The rain continued in a cataract, pelting and stinging my face with the force of a hundred little pebbles. We began to feel a growing sense of dissolution, as if the very spot we stood upon–where glossy puddles began to circle our feet–was about to disintegrate. I turned toward the hills at the back of the ranch. They appeared to be melting as the waters poured over them and the earth disappeared beneath the washing. I could hear trees breaking, uprooting, and tearing from the ground. It seemed as if Nature were suddenly against us, mounting in thunder bent solely upon our destruction.

"*Calm down*!" Ralph shouted. "You, you, and you," he pointed, "start hauling all the transfer cages to the highest ground and begin making enclosures of anything you come across.

Everybody started to move, responding to the commanding orders.

"We'll split up into three units and begin evacuating the areas that are most endangered. Come on, let's get to the tack room."

Followed by dazed people, he passed out cat chains, halters, and lead ropes and began issuing new commands.

"Get every animal you can! Bring them up here as quickly as possible! Tie them to the semi or trailers or whatever you can find! If you can't catch the hoofstock on the veldt, cut the fence and run them up the hill!"

He split us up into four groups of three and four each, and we all began to function. And so we entered the approaching disaster with an exertion that wouldn't collapse until it was over.

Ralph and I hopped in Old Bessie and sloshed through the brackish, oozing foot of climbing water, making our way past the sound stage to the Junior Zoo. Bushes and branches began to twist in the wind, swirling down the same black sea we were now spinning in. "Get out of the truck, Toni. This is as far as she goes!" (I'd never thought of Bessie as a truck till then.)

You could see the ranch begin to separate as the waters cut loose from the mainstream and began to knife their way through the center of the park. The faster we ran the faster the waters rose. By the time we reached the Junior Zoo (just fifty feet from where we had parked), we had submerged two feet.

As we scaled the little hill and began unlocking and opening cages, we could see the mound breaking away on the far side. As the hill was being undermined, I made my way to the center cages to Yen and Yang. My shoes had come off in the thick slosh and were buried at sea. Ralph had the 500-pound lion, Major, by the leash. I had both leopards on chains, one in each hand. Carl Thompson (our only black trainer) was pulling Sultan and the other jaguars out. Someone threw a rope across the now three feet of racing water, and Carl secured it to the Arena, then pushed me ahead, saying, "You first." This seemed the wrong time for etiquette, but the leopards and I hit the current. If it hadn't been for the rope clipped to my belt by a large double snap, we would have been swept downstream. I hung on to the leopards' leashes as we made our way across in a kind of metaphysical occurrence: Yen, Yang, and I as center balance. I felt as though both my arms were about to be torn off. As a sudden onrush swept over me, I thought I would drown, but someone's arms were around my waist, pulling the three of us in. It was my "forever hero," the ex-con. Coughing and choking, I climbed with the bewildered, wet cats up the steep hill behind the wild section and clipped them to a tree above the railroad tracks, then slid down and went back for Kasan and whoever was left. As I entered the river, Ralph was having a difficult time with the panic-stricken Major, who did not want to leave his familiar surroundings. The lion had pulled Ralph into the waters, and as they entered, the utility pole near the offices snapped in two. Hot wires exploded around them, and the pole fell near them. I saw my husband and the lion bobbing up and down. Thank heaven he had the lion on the end of that leash, for Major, in his stricken state, pulled himself and Ralph out of the torrent and onto momentarily safe ground.

The Junior Zoo was an island now, and the only animals left were the two wolves. Since I was the last one across, I undid the ropes from the arena and tied it around my middle, and once again someone reeled us in. As I came across the second time, I had the sensation of not being able to distinguish water from air; I was taking in as much of one as the other, gagging in a liquid suffocation.

This time it was Ralph who pulled me from the torrent. Someone grabbed the wolves from me and ran up the hill toward the upper parking lot. Carl, Ralph, and I bent over, holding ourselves by our knees and breathing in big, heavy gasps. Then came the most blood-chilling noise I ever hope to hear: crunching steel as it ripped and tore apart as part of the sound stage came crashing down. Trees, logs, and a smashed, gnarled pile of tin in great waves of arching water collided with what was left of the little hill at the Junior Zoo, and buried it forever before our eyes. With those obstacles out of the way, the waters from the channel and those racing through our park joined.

I was now beginning to feel fits of hysteria. As we began unlocking the cages in the wild string, a camel, then a giraffe tumbled by. We could see they were dead and were now a part of that awful, churning debris. I managed to open five locks and fling back the doors, allowing those animals to escape. They splashed in heaving leaps through the waters, up the steep escarpment, and out of sight. Marty had arrived upon the scene and had begun to shoot tranquilizer darts into the really wild animals who lived here, popping them one at a time as he went down the line. Carl lassoed a lioness in her cage, and when he pulled her out, she went completely insane, bolting and running with him into the swell. They disappeared before I knew what happened. I looked around for Ralph, but he was gone. The waters where I stood were now at least three feet deep.

I reached the bobcats' cage and just as I did, I dropped my key! "No! Please God, no! I dropped my key into the water!" I couldn't get the bobcats out. I shook and pulled and tried to undo the bolts that were holding the sections together. Timmy and Bobby, the two lynx I had raised from babies, were on top of their den box, looking to me for help. When I look back to that day, I still see those amber eyes beseeching me. A wave of defeat overcame me, and I shrank from this unbelievable destruction. I wanted to die with them. I just wanted to die with them.

I began screaming down the row, "*Get them out of there! Dear God, get the animals out of there! I don't have a key!*" No one heard me because no one was there.

Then someone yelled from the top of the escarpment. I looked up and saw Charlie and Marvin on their way down. *Guns* were in their hands.

I stared up at the two men in rising disbelief as they waved across the river for me to get out of the way. The raging waters were deafening, and the vast sheets of gloomy grey rain made it impossible for anyone to see. My frantic hand signals meant nothing to them. I pressed my fingers through the chain link and stroked the wet, warm fur of Timmy and Bobby—for the last time.

The end of the world came with a frightful explosion as what remained of the sound stage fell and was mashed like so much scrap metal—and so became part of the dead sea. Those remaining walls had been responsible for rechanneling a great part of the water. When they tumbled down, a fourteen-foot tidal wave filled with steel beams, logs, and sets suddenly moved toward me. I could hear the rifles shooting, killing our animals, and I wondered if there was a bullet for me. I began to scream in futile anger, shrieking at the top of my lungs—until those waters came crashing down with terrible swiftness.

I must have bounced off the cage and slammed against the hill, for the next thing I remember was pulling myself up the embankment and suddenly clinging to Yen, who tolerated my sinking my nails into him, pounding his side, and sobbing into the folds of his neck as tears and rain and rivers swept away so much of what I had come to know and love. In my crazed mental state, I had the overwhelming feeling of wanting to die. I felt I should have gone first to clear the way and make all the proper arrangements for my animals. I hurt so deeply, and I was having great reservations about a God who could let such a terrible thing happen.

The two leopards and I climbed as far as the railroad tracks. I did not look back—not even to see if they were all dead. I didn't have to. As we walked, I was haunted by the memories of the last few moments. Those haunting memories remained with me a long while afterward.

My mind then became inactive; I followed the rails in a kind of somnambulism, the rail still pouring, the devastating flood still overflowing. As I headed for the upper parking lot with the cats, a Rolls Royce Silver Cloud came up behind us. The tires of its luxurious wheels had been removed and the naked rims clung to the steel rails. Gardner McCay had arrived. P. T. Barnum would have been envious of that entrance. The sight was so outrageously out of context with all that was happening that it brought on a new convulsion of hysteria in me.

We now went past the tree where Scalla was buried and made our way down to the parking lot, which, in my absence, had become the center of bustling activity. I handed the leashes of Yen and Yang to someone I had never seen before. The man took the two cats from me as if he had known them for years, and they vanished into the racing activity. I shook myself awake and, seized by fear, ran toward the nursery. Thank God; the building was still there. I was about to wade into the turbulence when Stevie came up behind me and said, "They're all safe; don't worry. We were able to get them all out before the flood hit."

"Even the snakes?" I whined.

"Even the snakes." She smiled. "We put everything in the Duka."

I went weak with relief. "Have you seen Ralph?"

"I think he's still in the parking lot," she replied.

"And what about Carl?" I asked. "He was swept away right in front of me."

"He's all right, he wasn't hurt, but–" She hesitated. "We think Bert may be dead."

It seemed as if we were caught up in someone else's insanity, and although each of us wanted to stop and talk–to try to understand and make some kind of sense from the monstrous devastation, we found we were so much a part of the crisis that neither could help the other cleanse her mind of what it had seen. With bewildered frustration I went back to the parking lot. The sky in spots was beginning to clear, and as the drizzle stopped, through the smoky mist pale streaks of shining light beamed down upon our overwhelming misfortune with almost a mocking, celestial delight.

Through the hazy overcast I could see Ralph standing in inexpressible sadness, slumping in heartache and despair. He was looking out, over all we had lost, over our sea of common ruin. I took my place beside him, and he took my hand in his and squeezed it tightly, then pulled me to him and hugged me with sorrowful force. Here in the arms of this man and with the life left in me, I began to cry.

We clung like that for a long, comforting time. Then, wiping the grief from his face, he said, "Enough of this! Let's get to work."

So many wonderful people had come to help, many of whom risked their own lives to save an animal's life, people we'd never seen before and were never to see again. They had walked in from the Acton side on the railroad tracks, having heard about our state of emergency on the radio news flash. And my dear ex-con, who had driven to the ranch before the entrance was washed away, called for the help of his friends over his Citizen Band radio when the phones went out. They had set up relay stations across the channel and organized supplies and emergency aid. All the while I had been at the Junior Zoo and the Wild String, these strangers, who had seen wild animals only in the zoo, were evacuating lions and tigers and everything imaginable, handling them as if they were their own pet dog and cat. It was one of the most fantastic events I will ever see, and if anyone ever needs proof of Affection Training, they have but to know of that remarkable day. People of all kinds had come, from teenage boys, who held a camel in one hand and a buffalo in the other, to little old ladies in levis rescuing monkeys and birds. In the midst of this heartbreaking disaster a powerful feeling of comradeship brought us all together. I wanted to hug and kiss every last one of those people I didn't know. They were one of the few triumphs of that day.

The flood remained in full sail for nearly an hour. It covered a span of some 500 feet, reaching all the way from Soledad Canyon to only thirty feet from the railroad track, destroying for the most part everything that had been in its path. All that could be done had been done, and until the waters lowered we could not prepare to resume our life. We made the animals as comfortable as possible and did the best we could to console one another, thanking each other in a pleasant kind of way for having been able to save what we had. That consolation raised our spirits somewhat, and even though the strangers were counting their own blessings, that did not allow them any less sympathy for our personal grief.

There came one glad moment, when Bert was brought in, drenched and spitting up water, looking a great deal like someone who was not sure he was glad to be alive. He had managed to hang on to a tree until the torrent had died down enough that he could be freed.

We were sitting in the Duka when someone handed us a cup of soup. It was scalding, but Ralph drank his down as he would a cold glass of milk, then resumed staring at the floor. There must have been thirty or forty people in the Duka with us, but none of them came over or said so much as one word, in their own ways empathetic to what we were suffering. Ralph stood and began pacing back and forth, back and forth, contemplating his destiny, mumbling incoherently, holding only himself to blame for what had happened. His undeserved guilt-ridden concern gave me a little renewal of strength, and I tried to come out of my own misery long enough to comfort him. But before I could speak, someone shouted that the waters were subsiding, and we could now walk out into portions of the sinking river.

It was time to enlarge upon our emotional repertoire.

Some time after the tragedy, as I read Wordsworth's "Excursion," these words raised themselves from the page as if the poet had written them after witnessing our ordeal:

And when the stream
Which overflowed the soul was passed away
A consciousness remained that it had left,
Deposited upon the silent shore
Of memory, images and precious thoughts
That shall not die and cannot be destroyed.

We followed the tracks past what used to be Beaver Dam to the far end of our property. Here we began our long hike back, stopping now

Little remained of anything.

The remainder of the Wild String.

and then to pull from the muddy trench something of our loss. But there was little left of anything. The hoofstock section lay in shambles, and as yet we hadn't taken inventory of which animals remained alive and which had perished. Inspecting the pens for what may or may not have remained was traumatic. Ralph and I found ourselves praying as we cut through the chain link fences—praying to any god that saw fit to listen. Hadn't we been crushed sufficiently? Hadn't we suffered ourselves sterilely clean? What more could possibly be accomplished by more grief? In our soul-searching we began to reconcile ourselves to a dismal view. Had we lived by the premise that we were in control of our own destiny? Had we lost ourselves in our own interpretation of how our lives should be spent? Did we need to be humbled by such an act of God? To have worked so long, so hard, was this a just reward for our devotion? Life, it would seem, had now untied our bonds of illusion.

Still, that only twenty animals had died out of 1,500 was in itself a miracle wrought by the laws of nature and should be viewed as such. For that consolation we were grateful, but why such a thing should happen at all bewildered us.

The majority of the animals that had perished were those in the Wild String. Their pitiful bodies were carried from the swamp that had once been their home and were placed on canvas wraps inside the one remaining building in the area, the mechanics' garage. I pulled my Kenya from the marsh and cleaned him off before laying him to rest. He had been shot in the head. Was it better that he had died in such a way? I suppose a well-placed bullet was preferable to drowning. But why did they have to die in such an offensive way? Why, God?

As that day progressed, a kind of mania presented itself in the form of a morbid public curiosity, and thieves and parasites made their outrageous appearance, stealing everything in sight while pretending to help us in our hour of need. But the worst crime they were to commit was not so much the pilfering of material goods; they violated a moral code as they stole even the dead bodies of our animals. It began to seem then that even humanity was against us. No matter where we turned, we found some new madness confronting us. It became increasingly difficult to cope. I felt drained, almost like the land on which I stood; all life seemed to have vanished from me. At our ranch, wandering among all I had known, I was as lost as anyone could be. Such an experience tends to leave you vacant. When a spirit finally moves back in, it is not the same. You will never be the same again.

I watched Ralph trying to control the mountain of confusion, somewhat caught up in it himself. But never did a man fight so bravely against such unwarranted hostility. That day was the beginning of his Armageddon, and it became a seven-year war of courageous Helfer restoration.

Indeed, a lot of ugliness followed. A major municipal zoo captured front-page publicity by announcing (by someone not in authority) that they had offered shelter to our animals until such time as we could get back on our feet. Stevie called from Acton to accept their kind generosity, and they flatly refused aid.

We spent three days and nights building new shelters and trying to protect what was left of the ranch in case the rains struck again. We began to police the property, keeping out not only thieves but souvenir seekers. Some of our own employees, through utter exhaustion and concern for the future, began to experience a sense of panic, as if the end of the world were very near.

When we began rounding up the animals that had been turned loose during the flood, it was amazingly easy. None had strayed far from home; some had wandered back on their own, and they accepted the new environment without hesitation, with more calm than any of the people around them. They went into their new enclosures as I'm sure Noah would have walked them into the ark. That their own familiar dwellings were not around did not upset them. In their faithfulness they returned, one after another, hundreds of them. One found his way back in the middle of the night. He was an animal from the Bronze Age whose name was Fearless, our cape buffalo. He made his way through the heap of salvaged cages and began to pound with fierce intensity with his bosk upon the trailer of our watchman. It was the only way he knew how to let it be known he was back, asking where to go. Why the watchman shot the buffalo is something I don't know. Answering the buffalo's call, he stepped out and shot Fearless five times in the back, and even though Fearless fell, he was too tough to die. By the time I came upon that insanity, Marty and Ralph were patching up the buffalo and had bedded him down just in front of the maintenance room. My head felt as if it would burst with this senseless, never-ending destruction. I wanted to be rocked and held—to be a little girl again for a while so I wouldn't have to think about the awfulness everywhere I turned. It struck me then there was no escape. Fearless had been given a sedative and lay resting in someone's protective lap. The gun was taken from the frightened watchman, who was not supposed to carry one or even have one on the property.

It was perhaps twelve or one when I got up from my sleeping bag in the Duka and made my nocturnal way down the tracks toward Gloria's trailer. (Gloria had separated from Bert after many years and was now living in a trailer parked next to the hospital.) Perhaps if I had another woman to talk to, I could find some perspective. I could see a curious glow, and I heard muffled voices as I came off the trestle. I switched off the flashlight and silently made my way in the dark toward the forms huddled tightly together around the fire. In the light they looked like ghouls in grey shrouds, attending a witch's sabbath. They appeared at this late hour to be having a barbecue. I stepped closer and found to my horror they were roasting over the flames portions of Enod, my beloved giraffe. In their depravity they were eating Enod. Her delicate head had been hacked off and lay to the side of them. Her skin had been stripped and was stretched next to it. I ran for Gloria's trailer, where I threw up.

When the sun rose I left. We had stayed up all night talking, evading the subject of the flood. She did not want to hear about it. Where I had hoped to find a sheltering mother, I found again that aging child of mute reality.

No one since has had much to say about that time. It was too dreadful to remember and too painful to dwell upon. This is the first of that past to be written down, the first calling to mind.

The disaster was claimed an act of God, for which there was no insurance. Our financial loss exceeded $1,000,000; our emotional loss could not be measured. Oddly enough the *Daktari* house had withstood the effects of the catastrophe–only to burn to the ground two nights later. Brave Bess and many other vehicles had perished in the storm. Mo pulled their crushed bodies from the ruins. The sturdy cement and steel rhino barn had partially collapsed with the rhinos still in it, but they emerged unharmed–miraculously–for the mud was up to their horns. It tooks hours to dig them out. It seemed to me that we were chiseling away at solid stone, like the sculptor of Rushmore molding those famous heads; for when the rhinos, Click-click and Luna, were freed, they were even then more of the ground than any boulder. It was as if our earthen statuary had suddenly come to life, a bit too soon perhaps, for they were still a little rough around the edges.

The three hippos were found a mile downstream having a perfectly marvelous time playing in the remaining slush. It had been fair weather for the cast-iron hippos. Alligators were discovered up and down the coast for days to come, their stone-age driftwood having washed all the way to the Pacific Ocean. It must have terrified those residents to think creatures such as these were also inhabitants of the sea. It was probably some time before any of them swam in that ocean again. If my memory serves me right, all but two of the alligators were dead. All we had lost were eventually found.

Humor has a way of popping up at the most unexpected times. Rever-

In better days: Ralph riding Click-Click the rhino; Luna in background.

end Dooley, our neighbor, who by some good fortune had not lost his house nor even a fragment of the surrounding junk yard, sent word to us by way of a volunteer he had hailed walking down the tracks. Beneath a pile of salvaged cars, next to his pig pen, a large cat lay—a mountain lion, to be exact, and would we please send someone to recover him, for the pigs as well as Reverend Dooley were not keen on having him there.

We were still missing several animals who had been turned loose—a mountain lion for one and some hooved stock. Most of the trainers were away rounding them up when the message came down. Marvin and I made our way to the neighboring rubbish farm, crawled under the mountain of cars, succeeding in putting a leash around the snarling cougar's neck, and pulled him out from under the man-made wreck with more than a little trouble. We attributed his panicked state to the disaster. Because of his alarm, we put two leashes around his neck, Marvin holding him on the one side and I on the other. In this way we insured our own safe return, for the cat was trying his best to bounce himself off us. We managed through his thrashing, slashing, hissing, and spitting to get him to the compound—whereon we were told our missing mountain lion had been found—and who was that we were holding? We had apparently captured a wild mountain lion! You can imagine the looks hurled from Marvin to me to the cat. We didn't think the incident very funny just then. All we were concerned with was who was going to let go of the leash first.

The saga of the wild lion had a happy ending. After a brief stay with us, and once we had all our own animals returned, he was taken back into the hills and there released into familiar surroundings.

As far back as I can remember one male actor, in my opinion, has outshown all others. Marlon Brando was one of the rare human beings I had always hoped to meet. Of all times, he chose to show up in my life just after the flood. Oh, hello, have you come to take me away from all this?

At any other time I would probably have been shocked by his presence; instead, he was shocked by mine. I resembled the remains of something vaguely human. I showed him around what was left of our place with less than my usual Red Carpet flair. He had not come to view our demise; ironically, he had come as a result of a business arrangement with Ralph to stock with game his own island paradise near Tahiti. As he looked around at the devastation, he said sympathetically, "I'm sorry. I'm so sorry." To which I responded, "It's a shame you didn't come to visit last week. If you had only seen us then, you would have come upon our paradise." (I saw it still.)

We moved that year to higher ground. Africa USA, even in her fallen condition, was very hard to leave. All my world, outside my parents' house, was here—everything I knew. It seemed very sad that our lovely

Eden now lay face down in the mud. To say goodbye to the land you love is a hard thing to do. My friends were buried beneath her; they were now part of this long stretch of earth, more a part of it than I would ever be—deeply rooted.

Two weeks after the flood Ralph became very ill. Night after night he woke gagging and choking as if he were drowning in his sleep. He spent some time in the hospital. Yet I knew his sickness was not of the body. His heart had been broken, and it would take a long while for him to radiate that strong vitality again. He sat for long periods looking out the window of his room, exhausting himself with what he saw, not saying a word, as if only the shell of him remained and somewhere along the way the soul had been lost. He appeared to be in mourning. If I made an effort to comfort him, he asked me not to speak. So I spent that time near his bed thinking about things I was missing. I had a difficult time accepting the fact that Ralph was human; it set me back somewhat. In a way, though, I loved him more for it, because I was not now held in such spellbound awe. It was equally wonderful to be in love with a *man.* I could help restore a man, whereas one on a higher plane would not have needed me.

Having survived the worst times of our lives, we would build on the past and create anew.... We moved that year and began all over again. And thank God we had the opportunity to do so.

That, as I said, was in 1969. We still have our animals; they continue to star in the movies and on TV. Our new compound is The Gentle Jungle, appropriately named for its inhabitants, who are, as before, the products of Ralph's and my affection training.

And we still teach many students who come to learn our method—to go out and teach it in turn. We are hopeful of its widespread results.

Still, that move from Africa USA marked the end of an era. I am glad that at last I have found the courage to write this chapter. Perhaps by not speaking about the flood for so long we had hoped to kill the memory of the event. But I cannot now, nor shall I ever forget, that once there was a land—such a splendid place on earth—known to some as Africa USA; but to Ralph and me that glorious spot shall always remain our Shangri-La.

Vignettes

Now I sit among wild flowers
Fur pettles, ivory stems, thorns of claw.
Flesh begins to show through sparse hair
As pelage adheres to me and starts to grow.

–T.H.

These brief anecdotes continue the book's philosophy of affection toward animals and reveal more about animal friends the author has introduced in this book.

–The Editors

Irresponsible

Upon ridding himself of the greater portion of his menagerie, a trainer brought to us a handsome lion. While this man was around, the lion was happy, but when the man left, the lion wailed, and no expression of understanding on our part soothed him. He did not know us and did not want our attentions. It was the man he longed for, the man who had deserted him and left him to pine.

The lion waited with faithful, unnerving hope for the trainer to take him home. He looked beyond us to the spot his friend had departed from. He ate sparsely and slept lightly, expecting each footstep to be that of his undeserving excomrade.

Ralph called the man and said, "Your lion is pining pitifully for you. You must come and take him home." And the man said coldly, "I will not; he is yours now. I am under no obligation." And Ralph said, "But you are obligated to those who love you; you cannot toss aside their emotions. You have made him care deeply for you, and you are responsible for his feelings." The man said, "I am responsible for no such thing. He is a lion, not a human being, and he is no longer my lion; he is yours." At which point he hung up the phone. And the lion died.

A Controversial Woman

One Saturday afternoon as I was shop hopping in Los Angeles, I came across an African gift boutique and was pleased to find an assortment of wildlife conservation-antipoaching-antihunting material. Everything was neatly placed on top of several glass counters.

The owner of the store came out from behind a desk in the back corner. We exchanged greetings and I told her I thought it was admirable for a business woman to be promoting such controversial material. She smiled and thanked me. As I was having a look around, I came across a good-sized display of elephant-hair bracelets and a wide variety of ivory carvings. These things were so completely out of context that I could not refrain from saying, "I presumed from the advertisements that you were against hunting and poaching and for wildlife preservation."

"Oh, quite so," she said. "I most certainly am."

"Well, then, why are you in possession of elephant hair and ivory?"

"I don't understand what you are referring to," she said with a puzzled look upon her face. "You must be under the misguided impression that animals are in some way hurt in order to obtain these bangles."

"What an extraordinary comment. Who told you otherwise?" I inquired.

"Why, the man I buy them from. He lives in Kenya, my dear, and is a well-known authority on wildlife," she added with pride, very impressed with herself for knowing a Kenyan.

"And exactly what was it he told you?" I asked.

"Oh, it's quite simple," she began. "When the elephant is rubbing contentedly against his tree, an African quietly sneaks up behind the tree trunk, grabs the elephant's tail while he is preoccupied with scratching, and quickly the man snips off the hair. Within a few months the elephant grows new hair, and everyone is happy."

Exercising considerable control, I then asked, "And how did he tell you the ivory was taken?"

"Well, that is even easier. These are baby tusks; they drop out on their own when the elephant is seven or eight years old, making room for the big tusks to come in."–

A Feathered Ball

During one of my more unique exploits, as I wandered in the turbulent wind, walking in pleasant harmony among the electrical forces, a tiny feathered ball fell upon the ground before me, nearly landing on my head as he was blown down from a tree. Had someone called, "Here, catch," I could have caught him. Instead, he lay upon the dusty road. Compassionately I scooped him up. He was stunned by the fall. "Well, no wonder. Tiny baby bird, never having flown, isolated now from his nest by a forceful gale."

Everyone said, "He won't make it. His age is against him; he can't live." But I took the little woodpecker home and changed his destiny. Somehow, with the mealyworms and the care I gave him, he thrived. He grew into a beautiful bird, richly ornamented with bright, flaming red hair and a slick, shining black feather coat.

Woody rode on my shoulder like a stuffed ornament. On this choice spot, he would suddenly spread his wings, startling people. He drilled on command. I had only to say, "Woody, attack," and, though I've flown large birds of prey all hook-billed and strong, never did I know a hunter so skilled, so proficient in his technique. Under my direction he flew to protect me, urged by my "Let him have it." Then, after he had left a peck in a bewildered noggin, I'd call, "Woody, come here," and pat my shoulder and lean my head to one side as my little marksman returned. It was not unusual to see our friends appear for dinner in crash helmets, previous victims of fearful "Woody the pecker" and his infamous precision drill.

On the Move

I often leave the front door to my house open to allow the breeze to clear out the still air within. One crystal morning the shadow of a cat drifted past my door—a cat old as time, atrophied, disintegrating, wasting away. Nobly she entered my doorway with a dowager air as though it were her home. I saw a flame burning behind the oozing eyes that could no longer see in the night. Her heart had not grown old. She purred and spoke as though she knew me, then wandered through the rooms and settled comfortably on a sheepskin rug; dignity reigned. Tiny free spirit, take courage. You are alive though dying.

I examined her and found her deaf when I snapped my fingers. I stroked her mud-caked, matted fur—no phosphorescence here. And I discovered the deterioration of her bones. No teeth remained in her mouth; her gums were bare. She cried in a faint little meow, then curled up in my lap and died.

The Conquest

In my nature is a deep love and appreciation for elderly animals. They are an experienced race. Age appears to have brought virtue to all their sins, and God pardons misdeeds by giving them a dignified serenity. Tranquil from having lived through such a time, they assume an agelessness and never seem to grow old but instead to grow conscious, having achieved liberation from their own timidity–a noble conquest by antiquity.

I suffered a long time from pains in the heart for animal friends of mine, and thought, without the time to exercise it, of rallying against the prejudice. In little ways I defended the unspoken argument but never stated it loudly enough to ease my guilty mind.

I have known animal actors and actresses who have given performances worthy of Academy nomination. Several in particular, who, through the years, have been acclaimed by critics and subjected to overwhelming public adoration, have remained modest and humble during their resplendent shine in stardom. And they worked, for the most part, because they loved their profession and all those associated with it.

But the profession prefers not to recognize these particular entertainers. They remain an anonymous group, never having been so much as acknowledged by the Emmys or the Academy Awards. (Believe me, Hollywood, you could not do without them.) So, isolated by their peers, they continue to excel in spite of persecution and imperious bias.

They do not carry SAE or SEG or AFTRA cards, nor do they receive residuals or social security, nor benefits of any kind from the profession they have served so well. When they retire, there is no movie home for them to recline in, no residence for these aged actors who made millions for Warners and Disney and MGM and 20th and Paramount and all the rest of them. They were an integral part of the birth of my town, and their ability to save a mediocre script resulted in the studio's financial gain. And for this they were held in abeyance. But not one fell by suicide or drug addiction or extreme depression or mental disease. So much of me was in these actors that I took personal affront at the industry's shameful, delinquent neglect.

Remember for a moment my beautiful King, better known to the screen as Fury. The studios who owned him "way back when" sold him to the highest bidder when he had done sufficiently right by them. Faded glories resulted in that magnificent actor's sale into 20th-Century servitude (they changed his name so no one would know those righteous, imageable men treated their stars so). His spirit as a reigning star was such that ideological years separated him from other equines. For ten years after his last great part he suffered horrible, agonizing abuse. Ten years of going to the auction, of living in pig pens and behind chicken coops, of encounters with whips–of being strapped in harness, pulling a farmer's rig–of fighting gangs of unworthy opponents–all the while maintaining such pride. Through ten long years of shattered expectations, of broken dreams and bewildering heartbreak he suffered, up to what could have been the last day of his noble life.

Come on, Hollywood, you're bigger than that. Show us things have changed. Give a little credit to a deserving minority race. Honor *all* your own. Then someday build a community where aged animal actors can retire to a place where the public and fans can come. You will be remembered and applauded for such a gracious cause. And through this act of kindness, you will be the Hollywood I have known.

December 18, 1967

My Dearest Toni:

I know how great the love you felt for King and he for you. It was probably the strongest feeling you have ever had for life.

That makes your loss very deep, but the fusion between you was such that he has entered you in spirit, and although you will miss his physical presence, you and he have become one. King will be preserved by your love, and the image of him will remain through you. So when the hurt has lessened, write about him and paint his portrait and keep his memory alive. For the two of you together had reached perfection, and the world stopped when you came upon it, and everyone, allowed to feel your love, was momentarily caught in your glow. They all knew then that an emotion so strong really existed. It gave others hope.

So please, don't bury it with King; feed it with the memory of the years you may never have had together, and let it grow with time as all great moments do.

Take courage and know that I love you in the same way.

Ralph

SINA

I have often felt that the only obvious mistake nature made was in the creation of man—man, made of many parts, some good, some bad, and some so completely idiotic that we have to laugh at them and not with them, lest we become as ludicrous as they.

Let me recall some neurotics from the human animal world for a moment and elaborate on the warped mind of an organization of these creatures, who in 1965 squandered (so I'm told) several hundred thousand dollars of inheritance in an embarrassing attempt to "clothe our animal population."

We fell under severe attack when a group known as SINA (Society for Indecency of Naked Animals) concluded its investigation of animal nudity existing at our compound and declared the entire region a "moral disaster area." By post we were informed that "our exotic animals had been observed in a completely unclothed condition, and if we were decent citizens we would keep their nude bodies off the streets and out of the studios until they were properly covered."

SINA then issued a statement to the press, saying that they were aiming this campaign at quadrupeds higher than four inches. "When we have these animals decently dressed," a spokesman said to a reporter, "then we can start doing missionary work among the wild." The society suggested to us team uniforms so we might identify species more easily. (I did not recall having a problem telling the difference before.)

A self-proclaimed New York fashion expert was to drop from a plane a complete wardrobe designed especially for our animals, and to expedite matters in case he was to miss target, an ambulance would be on hand to pick up the clothing and rush it to our critical location.

Ralph fought back with overwhelming cooperation. Having the good taste to possess tact in any situation, he offered unlimited help when it came to dressing the primates, but he pointedly informed the SINA agents that when it came to the lions, tigers, leopards, wolves, hyenas, hippos, rhinos, giraffes, and snakes, they were on their own. (Oh hello, here are your clothes; please crawl in.) And, of course, since their mission of good will provided us with this generous supply of garments, he took for granted that they would also assume the laundry service.

After due consideration on their part, we received a brief but to the point phone call from SINA. The cultured voice expressed this message: "After much consideration and with great apprehension on the part of some of our members, the society has given your case special thought, and the conclusion we have reached is that under the circumstances we shall regard it as perfectly natural to leave your animals nude. Hereafter they shall be looked upon by us as works of art and will fall into the same category as Venus and David." It would seem, when all was said and done, our friends were to be judged as masterpieces—more than mortal flesh.

And the suspicious, prejudiced little minds went on to diaper the dairy industry.

Mad, you know!

Tana at 4

A small child of four-odd sat with Prince one day, shaking a long stem of fuzz loose from a seeding weed. She blew it to an Etesian breeze and, waving a fond goodbye, turned to her friend and said, "Listen, you don't fool me. I know you're not a dog." And he, with a tilt of his head and a half-cocked ear filled with silence, refused to talk about it.

She put her arm about his neck and gave him a brave hug as though she would keep secret this unknown thing, and while their minds were touching, she spoke of his heroism, and their gaze fell upon the rippling waters of Swan Lake.

The long tail wagged, and the pink tongue took pleasure in washing her rosy face, and she, wiping the moisture from her skin with the sleeve of her shirt, spoke of how together they would roam the globe.

The little person of four-odd was so perceptive, as though four short years had developed such an aware individualist that she knew the true nature of things, not by a mere venture into the imagination but by the experience derived from wise intelligence, of reasoning and deep deduction.

Round I

George was my very favorite tarantula. He was the mascot of the nursery crew. Every day one of us would take him out, set him on our sweater, and go about our business, wearing George as one would a living brooch. Although George was terribly healthy and always a hearty eater, I had nóted his sleek coat beginning to turn a somber shade that summer. During this time his appetite began to drop off considerably, and I became a bit concerned.

It was just a little past dawn when I opened the nursery to begin my day. Things were always a terrible mess in the morning. The tenants seemed to do their best to devastate their apartments each night. I was thoroughly convinced they worked so diligently at it simply to make my days (as their maid) as miserable as possible.

Because of my fret over George I went first to his glass case to inquire as to his state. To my surprise, there was George looking better than ever. His coat was gleaming beneath the fluorescent light. Next to him, to my absolute astonishment, lay, deader than a doornail, another tarantula, stiff and rigid from rigor mortis. I came unglued. The very idea that someone on our ranch was holding tarantula fights in the black of night in my nursery was enough to boil my blood. I saw red as George, his glass case, the deceased occupant, and I went to find the culprit who would do such a sick thing. (At least George had won.)

"Okay, which one of you guys is responsible for this?" I exclaimed, approaching one after the other. No one would 'fess up. "I'll get to the bottom of this, you know; there will be no peace around here."

One by one I encountered them. Either they were good liars or they were genuinely innocent, for each seemed equally as startled regarding the bout. I had advanced upon everyone, with no results. I had no choice. I would have to put this matter in RDH's hands. I didn't like to bother him with this kind of thing, but there was no alternative. So in a goose step, I marched myself past his secretary and into the office of the chairman of the board.

"Ralph! You will not believe what's happened." As I proceeded to tell him my story, a faint smile began to take its place in the corners of his mouth. "Why are you smiling?" I asked. "This is no laughing matter. I'm very upset, and I think you should be too."

"I'm sorry," he said patronizingly. "Please continue."

"All right, but try to take this more seriously," I commented. I continued with the saga of the gruesome fight and ended with the sad outcome of one dead victim. I thought I had expressed my case rather eloquently, but for some reason Ralph began to laugh, and he laughed; then tears started running down his cheeks as he laughed harder.

He wiped his eyes with a tissue, as I asked, "Are you quite through?" He nodded his head, but I could see he was on the verge of convulsions again. "What is so funny?" I asked. "I fail to see the humor in any of this."

"I'm sorry; it's all my fault. I should have told you before this," he mumbled between chuckles.

"What? What should you have told me?"

"It's the tarantula." He started to lose control again.

"What about the tarantula, Ralph?" I was becoming quite annoyed at his behavior.

"*They shed*!" He exploded before he lost all composure.

Old Mo

Move *up*, Mo!
The ragged, flapping ears slapped alert.
The massive trunk swayed, spraying dust.
Huge mammoth pondering feet stirred,
Keeping in step,
As though waiting for the inner beat.
Then, gathering it all together,
All parts lurching in unison
. . . it *moved*!
Nine thousand, one hundred eighty
pounds.
Seven feet, ten inches
. . . moved.

Move *up*, Mo!
Her tail swayed as if to keep the balance,
Her trunk touching out in front of her to
check the ground.

And if you fall, don't fret, Mo.
The chains, the cranes, our love will
get you up.
Trunk *up*, Mo!
Lift me up. (I need it now.)
Pack your trunk, smell the flowers,
Fall asleep against the old oak.
Talk to the birds, talk to the children,
Lift them to your mighty back.
Spray them with water till they giggle.
Your trunk was so long.
A trunk was born
And God added an elephant.

Lean way down; I'll put my eye to yours. Or is it yours? For it seems another is looking out at me from inside.

How you loved it when we patted your tongue and pulled your teats. The bellorous rumble emitting from your cavernous tum seemed to say, "I was great, wasn't I!" And you were! So great! So kind! So gentle!

A good life you had, Mo. Seventy-eight years. You were the oldest! World-traveled —India, Germany, England, the USA. The great days of the circus, the calliope, the midway. "Step right up and see the great Modoc!" Center ring you had, Mo. Remember the lights, the popcorn, the clowns. "Ladies and Gentlemen . . . children of all ages." The great tent would darken, voices would still, the music would begin. A spotlight blasted the darkness on you, Mo! Bedazzled with sequins, the spot would follow you, swaying and bowing. Trunk up! Head down! No trainer, Mo. You did it—alone, alone! Lights on! An avalanche of applause and then back to the menagerie tent. Pushing, pulling, helping—always helping, no complaining.

Remember the fire, Mo? The screaming

people? The frantic animals? How many you saved? Pulling circus wagons free, pushing, lifting fiery beams. And the poisoning? So many died, but you survived. Remember, Mo?

And then that horrible day the mad, drunk keeper blinded you in one eye. One-eyed Mo, they called you. Couldn't use you then. No green pastures, no thanks. For ten years you stayed in that zoo, Mo. So thin, so depressed. I didn't even have enough money to buy you. Had to borrow. Not even a truck to haul you. Took a loan. We suffered hard times together, Mo, good old Mo. I owe most all to you. It was you who gave to me! You trained me! You taught me how to "affection" train!

Move *up*, Mo!
  Move up through those great gates!
  Make way for the best of them all!
  Crank open those pearly gates just a
    little wider;
  Strengthen that bridge a little stronger
      for Mo's comin'!

Move *up,* Mo!
  Give her a railing to guide her,
  She only has one good eye.
  Call out to her loudly but with love;
  Her hearing's not too good.
I know her great legs are weak,
  But she can make it.
You don't have to worry about her
    complaining
  Or giving you any cause for trouble;
    Mo's a good girl.
    It's time to go, Mo.
    May you live in a gentle jungle.
  And when my time comes,
    Let down your trunk, Mo,
  And lift me up

—to *you*,Mo
—to you
—Ralph Helfer

The Haitian Tiger

We were to be on location on the remote island of St. Croix in the U.S. Virgin Islands shooting a film for American International Pictures. The movie, starring Burt Lancaster and Michael York, was called *The Island of Dr. Moreau.*

We sailed with twenty-six assorted animals from Miami on the freighter *Freschoki* and arrived in St. Croix a bright shade of pea green in the latter part of November. One of the first things to catch our eye in the local paper was a small, unobtrusive article announcing the intention of a karate expert to fight a wild tiger to the death in Haiti.

"What some people won't do for publicity," I thought. We gave it little thought and continued our hectic schedule—until the following week, when an official announcement was made that Haiti had in fact scheduled a fight between a tiger and a man. Apparently thousands and thousands of letters poured into the Haitian government house protesting the bizarre event, and because of the adverse publicity (and since Haiti was promoting tourism), the bout was publicly cancelled.

The next we read of it was in a news statement revealing the fight was to take place in Uganda. Now the detective work began. Haiti did not have a zoo—did they have a tiger? We were 500 miles from that land, and it was extremely difficult to receive any kind of reliable information. Several officials assured us that the tiger had arrived and had been sent back to the zoo in Japan whence it had come. On another occasion we were told that it had been shipped to Uganda (in which case there was nothing we could do, except verbally protest to the powers that be).

We remained content with the story of his returning to the Japanese zoo, until rumors came to us from Puerto Rico about a dying Bengal tiger held captive in an isolated warehouse on the far-away island.

At a cast party (a truly gala evening), we made arrangements to charter a twin-engine Beech aircraft, and romantically we prepared to kidnap the dying tiger. (That is, if we could find him.)

The plot thickens....

Late that night we received a call from an executive of the Humane Society of the United States who was in Haiti and had located the cat but was unable to get him off the island. At 4:00 a.m. that morning the pilot-copilot, one of our trainers, John Gillespie, and I left on our journey. It had seemed like such a good idea at the party, but when we were airborne, we had second thoughts. Neither John nor I had a passport. Our destination was under a dictatorial government, and we had no plan of action. This had all the markings of a bad James Bond movie.

We disembarked and with great arrogance marched through customs as though we owned the place. An official stopped me and asked "What are you doing here?" "Oh, just the usual," I replied and walked on. He shrugged his shoulders and let me pass. When we found "our man in Haiti" we proceeded to the infamous warehouse.

I have never seen such security around a place. Guns were everywhere, mostly pointed in our direction. There were locks on top of locks. When the doors were finally opened, we saw a Royal Bengal tiger lying in a 3-x-6-foot shipping crate on a bed of his own feces. He had lived in that way for some twenty days. His muscles had atrophied, he was dehydrated, and from what I could tell he was blind in one eye and had problems seeing with the other. We were later to discover that the animal's head had been scored. All the martial arts expert would have had to do was hit him with one powerful blow, and the cat's skull would have cracked.

He puffed happily when we approached him in his confinement. He was not totally wild. We now became both determined and confident that we would be able to accomplish our mission.

Before they could change their minds, we loaded the crate and its contents onto our plane under armed escort, revved up the engines, and were about to take off when alarms and sirens sounded, and people with guns came running up the field in our direction, screaming incoherently in French. Terry took off anyway, and we screamed back, "You can't shoot us; we're Americans!" I was suddenly filled with red, white, and blue. That's what years of good brainwashing can do for you. That and a good case of *stupidity* can make you completely disregard self-preservation. (Being an American sometimes does strange things to your mind; imagine walking into a foreign government and telling them what *you* are going to do when you have absolutely no jurisdiction there!)

We were in the air! Heartily we congratulated one another on a job well done. "Mission Dying Tiger accomplished." What was the Raid on Entebbe compared to the Raid on Haiti? The tape had self-destructed!

Have you ever flown in confined quarters with twenty days of defecation? It didn't seem to matter that all the networks as well as UPI and many major newspaper reps were there cheering as our little Beech aircraft touched ground. By that time we were all in grave need of smelling salts.

We named the tiger *Freedom,* and when we left St. Croix, he left with us. Reporters came from far and near to cover his story, and all the way home across the U.S. we stopped in major cities and held press conferences about the kidnapping.

Somehow, when an article in *People* magazine came out, the rescue was largely credited to the Aga Khan, who was not in the least involved. He was also given complete financial credit as well as great moral support. So to keep the record straight . . . .

Your highness, if you should happen to read this small story, you owe me $3,000. You may keep the credit.

Yours truly.

P.S. Freedom is living happily ever after in the suburbs of the San Diego Zoo.

Tales for Tana

When Tana was three, I returned to the house after making a trip to the junction for supplies to hear what sounded like a class going on inside. I opened the front door, my arms filled with groceries, and saw Tana sitting crosslegged between Kong (the gorilla) and Hannibal (the orangutan). All eyes were on Ralph as he patiently taught them how to snap their fingers and pose as Hear No Evil, Speak No Evil, and See No Evil.

Their attention span was only momentary. Hannibal sat like a huge slob of a red frog, while Kong ran around and hit him on the top of his head, the way one would hit a scale at the fair to ring the bell. Each time Kong hit Hannibal with his closed fist, Hannibal sank lower and lower as though he were being pounded into the floor. Finally he lay flat out, and Kong rumbled and bent over him, poking him in his soft stomach to see whether or not he was dead. Hannibal suddenly reached up and, grabbing Kong by the ears, butted heads with him. Kong stumbled in a daze until Tana and Ralph joined in, and the four of them chased one another around the room and out the front door, then down the dusty trail to the nursery far below, racing in the fine breeze of a clear day. And Kong and Tana and Hannibal knocked on every door and asked if the animals could play, and could Daddy please, oh please, tell one of his lovely tales. And so beneath the learning tree we all sat as Ralph began:

"One by one, after having their naps of course, in leaps and bounds and hurdles and vaults they jumped one another all the way through the tall fields of green, where on a rise not too very near, butterflies stood in line by the tens of thousands, preparing to walk over the rainbow and slide into a pot of gold. Now, walking over a rainbow is a very delicate thing, for one has to be transparent, you know, and to splash about in a pool of gold all soft and yellow and resistant. You must be suggestive of having grown among the flowers and to be fully aware of the effects of such beauty, since when you emerge from having been wet by gold, a mysterious, overpowering quality lends itself to the enchantment; and anyone looking upon you is cast with rapture under a delightful spell–oh not a big spell, mind you, just a little magical one in which you must follow wherever they go, down the blossomy trail, skipping and whistling and dreaming all the way.

"And the golden butterflies, who by night are fallen stars in disguise, will shine brightly so that there is no end to day, and they will scatter glittering fairy dust to lighten the path, and the world will glow and gleam and change markedly in form so that everything will be covered in luminous brilliance, as though the sun had melted like lovely butter all over just about everything, except you, of course, because the stuff is really hard to get off. I mean, you know how gold sticks to some people's fingers. Why, there are those who, when they get near it, are never, ever the same. So I've made quite sure that it won't touch you in any way; but you may admire such a beautiful thing.

"Now, having skipped and sung and whistled all day, standing just beneath the huge arch of falling pastel mist, which, when it showers, covers you with a sleeping rainbow potion unless you sneeze, you could slumber for years at a time. And that would be awful because when you woke, everything would be the same, and you would have missed the whole story—'God bless you.'

"Now let's not be stopped by a flutter of flashing fairy wings. Lift up your fuzzy pants; you too, naked ape. And let's cross this paintbrush stream. Good. We made it. Here, let's stop for a brief exchange. Sulton and Riff and Penda and Raff, you are to change spots and stripes. Zamba and Tana will exchange manes, and Buda will give Kong his lovely red coat. And oh my, how silly you all look. Why, just take a peek at yourselves in the stream. But when they looked down, the earth, having been terribly thirsty that day, had gulped it all down, and so they could only look at one another. But that was only funny for a minute, since they could grow accustomed to anything, and if we aren't careful we'll bore ourselves and expect to be entertained by everyone but us.

"Now, all God's creatures, having lent a portion of themselves to their friends, have a whole new look, but that didn't change them inside; they're still the same. But if the tiger hopped like a kangaroo and if the kangaroo roared like a lion and if the lion walked like a gorilla and if the gorilla wasn't hairy like man, would that still make him a gorilla—?"

T. R. Helfer's Second African Journal

The Garden

Even a great artist could not have done justice to the garden encompassing our farm on Mt. Kenya, an old English garden that had been growing since the late 1920s. Fuchsias that had flourished for forty years or more covered the walls of the house with millions of lovely little bells, and humming-birds thrived on their succulence. Bouganvillea crept in flame across the roof to envelope it in rich color. Daisies, lilies, geraniums, and wild orchids with scents that could overcome the senses flourished at our house on the mountain, and it supported a large amount of life; its visual beauty was bestowed on anyone fortunate enough to behold it.

One January Ralph and I came in from safari with a group of sixteen people;

proudly we invited them to the farm to show it off. Villainous savage faces ran at us; and as we waved, they turned into smiling children, and open arms greeted us on our return.

As we walked onto our spacious front lawn, to our horror we viewed a flowerless, bushless, plantless, *bald* foreground. We stood speechless among the remains of splendor.

"Wahome! Wahome! What's become of the garden? What happened while we were gone?"

"Well, Memsab, while we were cleaning beneath the geraniums one morning we came across five snakes. Can you imagine, Memsab? Five snakes!"

"And did you kill them, Wahome?"

"Oh no, Memsab, we know how you and the Bwana feel about that. Instead we took out the garden so they would have no place to live."

How does one argue with logic such as that? You can't. And so without so much as a statement, we go on.

The Warrior

On my way from Amboselli one breathtaking day I stopped to picnic under a magnificent yellow fever tree in a small grassy glade surrounded by green palms and thick foliage–one that could easily accommodate a pride of lions. Impala grazed not twenty feet from where I sat; and as soon as I began to snack, vervet monkeys chattered down from the trees and scurried toward me in gleeful anticipation of a reward of food left from my lunch. Happily I tossed bits of fruit and bread their way and was completely engrossed in my fairy-tale scene when the monkeys turned and retreated as quickly as they had come. Back to their treehouse they fled. The impala stopped grazing, and the world around me became silent.

Thinking a predator was in the area, I packed quickly and was about to climb into the land cruiser when from out of the mossy forest walked a Masai Moran warrior. Spear in hand, he posed in storklike fashion, one leg tucked beneath his gown. The sun gleamed on his ebony skin in Rembrandt lighting. He stood proud and tall, neck extended, runner's legs elongated, copper curls strung long with string, and ivory bones piercing stretched ere lobes. Streaks of terra cotta decorated the stern face of the living statue. A forest of fever trees framed his aristocracy. The arrogant warrior stared toward me, and I was held in visual bondage. The contrast was such that it bordered on the ridiculous. I shut my eyes and opened them in case I had created this vision. For held in the strong hand of the mighty Moran was a pink-fringed umbrella, feminine and dainty in every feature, shading the exalted head.

To Please the Tourist

I was irreclaimably lost in the overwhelming beauty of the Mara, caught within the eye of Africa, wandering, my Rover and I, in wooded grassland among wild flowers and tall turf. The heavy rains of December had made this world lush with colored life. All dullness left my senses. I was vibrantly aware of my surroundings–at one with the world. I read the clouds as they migrated with me, listened to the invisible wind rocking the world, watched for signs given by birds in flight, signs of life and death. Defenseless I stood; passive I bared my throat.

Through waves of heat, 3,000 wildebeest drifted in floating motion upon the Loita Plains. The breeze of winged things fanned my damp skin where insects swam in salty pores. Fangs gleamed at me from under thick brush. The smell of my body grew strong from the touch of blazing sun. My lips were parched and cracked like waterless land–my barren eyes overflowed with new vision.

I passed a single Topi posed on a solitary hill. Elephants rocked beneath a canopy of acacia, sheltering their sensitive skin from the open sky. Powdered dust clothed their backs.

I came to the river, where flocks of guinea fowl scurried from my path. Hippos wallowed in baths of mud. The thick encrustation made them one with the soil, and they erupted on short legs to fall with the force of an avalanche over a massive boulder of grey. Whinnying, the river horse sliced the palpitating, murky sea and rested weightless.

My desire to discover an unexplored sphere was interrupted by the presence of a machine driven by an African. Goiter-eyed tourists filled the mini-bus. It chewed their pounds of flesh as it passed over corrugated ground. The driver herded before him a family of warthogs. He worked them as a cowboy does his cattle. The tourists now sagged from the windows, the results of over-indulgent lives. Their heads sprang through the roof; their cyclops eye screwed in, they salivated with lustful anticipation.

I could not see from where I had ascended why they should make sport with the little bald warthogs who ran, with banners flying, for sanctuary. Then I saw the shimmer of an amber eye, an animal still but for twitching muscles, frozen except for flickering tail. Now the driver ran them directly toward the amber eye, then backed away with his "shhh" of enclosure.

Blind, the warthogs headed for the lion, thinking they had escaped the jaws of the machine that was about to feed the lust of Japanese-photo invasion.

I could not be a witness to this intrusion upon nature, and I closed my eyes, hearing squeals and grunts and squeaks and screams, cries of the dying–then silence and the dragging of a dead weight. When I was able to see again I could hear only the tearing of flesh, the lapping of blood, and the crunch of bones.

Film rolling, cameras clicking, the machines had closed a curtain of steel around the enforced kill. It was the end of the play.

The Capture

A very good friend of ours and his wife were having a holiday in a rented cottage on the coastal beach at Mombasa. Since they were childless, their pet vervet monkey accompanied them everywhere. On one mischievous occasion the primate left his foster family's side and made a hasty retreat to a nearby tree, then climbed straight up to the top and, like the spoiled child he was, refused to come down. Instead he bombarded all with coconut bombs.

They called and they pleaded and offered scrumptious gifts. Still he held his station and was momentarily dependent on no one. For some hours he enjoyed his freedom, but for the remainder of the day he was looked upon as a serious problem. How would they get him down? His obstinancy and open rebellion had become an increasing irritation, as if before their very eyes the "darling" they loved most had turned into an uncontrollable monster. He had suddenly grown terribly unappealing, as well as foreign.

As it happened, a game catcher was nearby, and the couple called on him and several of his men to help them in their dilemma. The black game catcher asked that the couple not remain while he sought to capture the little monkey. They were told to return in two hours, and, feeling the situation was in capable hands, my friends left for the Arabic town to shop. In this way the men were allowed to go about their business undisturbed.

Two hours having passed, the couple on their return were met on the dirt road leading to the cottage by the smiling African captors, holding none other than the elusive little primate. So delighted were these people about getting their beloved pet back that they gave the men a handsome reward, put a leash around the vervet's waist, placed him in the car, and the three parted company from the compensated men. They were bewildered as to the way in which their friend had been recovered and discussed among themselves the possible methods. It was not for several hundred feet that they were to discover, to their horror, how it had been done. There before the once wooded cottage lay one dozen palm trees, scattered and fallen, having been cut and chopped to the ground.

Straw Flowers

My friend Mary had given me some pretty lilac straw flowers for my house. I arranged them in a great Indian copper pot and placed them on the mantel of the fireplace in my bedroom. They were just the touch of color the room needed, and it cheered up the stone grey wall of the hearth.

I took a short drive across the highway and up the Sirimon track to Mt. Kenya. When I came home, upon entering my room I missed the straw flowers. "Wahome!"

"Yes, Memsab?"

"What have you done with my flowers?"
"Why, I threw them away, Memsab; they were quite dead."
–It's Independence Day today.–

The Hunter

There lives in Los Angeles today an aging actor of worldwide fame, a man of considerable wealth who several years ago was searching for the ultimate in excitement. Many of his prosperous friends assured him that his thrill would come with the kill. And so with little time to spare between films, he elected to kill an elephant, the largest of God's land creatures. To insure success (since he had bragged of the experience he was about to embark upon), he called ahead to a well-known safari outfit in Nairobi. And since he could afford all that money could buy, he paid a considerable amount to have spotters locate the biggest and heaviest tusked bull elephant they could find. Planes flew high above the herds, and not until they had found the most magnificent of African pachyderms did the spotters separate him from his family and drop a dye marker upon him so he would stand out on the savannah. Then, equipped with sophisticated radio communication, beaters were landed, and they began to force him with their noise into the U-shaped valley beyond. On a ridge above the valley a tent camp was set up. When the actor arrived at the specified spot one week later by private plane he was accompanied by a full camera crew, who were to record his heroic moment for the distorted viewing by some TV watchers.

They descended to the valley, the most strenuous part of their journey a 100-foot hike. Armed with a high-powered elephant gun, shaded by an umbrella held by a young black boy, he sat leisurely sipping a Martini. At ease in these luxurious surroundings, he had his gun carefully mounted on a fixed tripod, the telescopic site adjusted to the moment. (With that rifle he could have hit the moon.)

He sat comfortably in that canvas chair discussing art and literature and creativity with the white hunter who was there to see that all went as prearranged. And as they were about to solve the problems of the universe, the head beater came into camp and announced to all, "Your elephant is coming just now, Bwana."

The beaters herding the mighty animal fell back as the cameras rolled and the elephant came into full view. Then, completely impersonal, and with no regard to the quality of life that stood before him, the actor set down his Martini to square the animal in the telescopic site, then pulled the trigger, as calmly as though he were throwing a dart at a circular board. A life that had graced the earth with grandeur came to an end and dropped to the earth with unexpected ease.

The film star removed the rifle from the tripod, and the two men posed with one foot resting upon the dead hulk as cameras registered and clicked away. Then a great round of applause and a cheer went up as falsely smiling, hypocritical men patted the actor on the back and shook his hand with gusto, congratulating him on his momentous accomplishment. They had a round of drinks and were about to become "masculinely drunk" when the skinners began to cut apart the grey colossus that "Jack the virile giant-killer" had brought down.

Our inflated actor jumped quickly to his feet, revealing that the sight of blood made him nauseated, and with the beat of a distant drum he boarded his waiting plane and made a hasty retreat back whence he had come. He soared with less than angels' wings, glorying in the power of having taken a life and feeling a tremendous sense of superiority.

The self-proclaimed demi-god expanded with satisfaction as far below in a valley of blood, skinners hacked away at ivory trophies, wove bracelets from tail hair, amputated strong legs to make waste baskets, tore ears loose to draw maps of Africa upon, and cut proud flesh to overlay attaché cases and shelter boots and fashion headbands–all to be sent to the States as fond remembrances of the condition of death and the "power" that had caused it.